MARE Publication Series

Volume 10

Series editors
Maarten Bavinck, University of Amsterdam, The Netherlands
J.M.Bavinck@uva.nl
Svein Jentoft, University of Tromsø, Norway
Svein.jentoft@uit.no

The MARE Publication Series is an initiative of the Centre for Maritime Research (MARE). MARE is an interdisciplinary social-science network devoted to studying the use and management of marine resources. It is based jointly at the University of Amsterdam and Wageningen University (www.marecentre.nl).

The MARE Publication Series addresses topics of contemporary relevance in the wide field of 'people and the sea'. It has a global scope and includes contributions from a wide range of social science disciplines as well as from applied sciences. Topics range from fisheries, to integrated management, coastal tourism, and environmental conservation. The series was previously hosted by Amsterdam University Press and joined Springer in 2011.

The MARE Publication Series is complemented by the Journal of Maritime Studies (MAST) and the biennial People and the Sea Conferences in Amsterdam.

More information about this series at http://www.springer.com/series/10413

Michael Gilek • Mikael Karlsson
Sebastian Linke • Katarzyna Smolarz
Editors

Environmental Governance of the Baltic Sea

Editors
Michael Gilek
School of Natural Sciences,
Technology and Environmental Studies
Södertörn University
Huddinge, Sweden

Mikael Karlsson
School of Natural Sciences,
Technology and Environmental Studies
Södertörn University
Huddinge, Sweden

Sebastian Linke
Department of Philosophy, Linguistics
and Theory of Science
University of Gothenburg
Göteborg, Sweden

Katarzyna Smolarz
Department of Marine Ecosystem
Functioning, Institute of Oceanography
University of Gdańsk
Gdynia, Poland

ISSN 2212-6260 ISSN 2212-6279 (electronic)
MARE Publication Series
ISBN 978-3-319-27005-0 ISBN 978-3-319-27006-7 (eBook)
DOI 10.1007/978-3-319-27006-7

Library of Congress Control Number: 2015960940

Springer Cham Heidelberg New York Dordrecht London

Printed on acid-free paper

Springer International Publishing AG Switzerland is part of Springer Science+Business Media (www.springer.com)

Preface

This volume presents research on current practices and challenges in the governance of the Baltic Sea marine environment – a complex and interdisciplinary research field of high academic and societal concern.

The book grew from the interdisciplinary RISKGOV[1] and COOP[2] projects on regional level environmental governance of the Baltic Sea, led by Michael Gilek and Björn Hassler from Södertörn University, respectively. These projects aimed to explore and compare arrangements and processes associated with the governance of large-scale environmental resources, problems and risks in the Baltic Sea.

We are very grateful for the generous financial support received from the Foundation for Baltic and East European Studies and to the funders of the BONUS+ programme (i.e. the European Community's Seventh Framework Programme and national funding agencies, the German Federal Ministry of Education and Research (BMBF), the Swedish Environmental Protection Agency, the Swedish Research Council FORMAS, the Polish Ministry of Science and Higher Education and the Academy of Finland). Without this support this book would not have been possible.

Special thanks to our fellow contributors, all of whom have submitted topical papers based on high-quality research. Finally, we are grateful for valuable comments received through the external peer review process as well as for the professionalism shown by our editors at Springer.

Michael Gilek, Södertörn University, Sweden (on behalf of the editors)

[1] During the period 2009–2015, RISKGOV (Environmental Risk Governance of the Baltic Sea) was an international interdisciplinary research programme focused on analysing regional environmental governance of the Baltic Sea. www.sh.se/riskgov

[2] The COOP project (Cooperating for Sustainable Regional Marine Governance) is funded by the Foundation for Baltic and East European Studies between 2012 and 2015. It aims to analyse and compare challenges and opportunities for cooperation in Baltic Sea fisheries and eutrophication governance.

Contents

List of Editors and Contributors

Editors

Michael Gilek School of Natural Sciences, Technology and Environmental Studies, Södertörn University, 14189 Huddinge, Sweden; michael.gilek@sh.se

Michael Gilek is professor in environmental science at Södertörn University and has extensive research experience in chemical pollution and other ecological risks in the aquatic environment, as well as in associated science-policy interactions. In his current research, MG leads international interdisciplinary studies analysing environmental governance and marine spatial planning in the Baltic Sea.

Mikael Karlsson School of Natural Sciences, Technology and Environmental Studies, Södertörn University, 14189 Huddinge, Sweden; mikael.karlsson@2050.se

Mikael Karlsson is a PhD in environmental and energy systems. His research focuses on environmental governance, including risk management, science-policy studies, chemicals legislation, climate policy and marine governance. Dr. Karlsson is also president of the European Environmental Bureau.

Sebastian Linke Department of Philosophy, Linguistics and Theory of Science, University of Gothenburg, Box 200, 405 30 Göteborg, Sweden; sebastian.linke@gu.se

Sebastian Linke is assistant professor in theory of science at the University of Gothenburg and studies the relationship between science, politics and society with a focus on fisheries and marine governance and the (changing) relations between scientists' and other stakeholders' knowledge.

Katarzyna Smolarz Department of Marine Ecosystem Functioning, Institute of Oceanography, University of Gdańsk, Al. Marszałka Piłsudskiego 46, 81-378 Gdynia, Poland; oceksm@univ.gda.pl

Katarzyna Smolarz is an assistant professor at the University of Gdańsk, Poland. Her research focuses on environmental exposure, ecotoxicology, risk assessment and combining environmental and social sciences for the protection of marine environments. In her ongoing research, KS works with endocrine-disrupting compounds and consequences of elevated CO2 levels in marine waters at different levels of biological organisation.

Contributors

Paulina Biskup Department of Marine Ecosystem Functioning, Institute of Oceanography, University of Gdańsk, Al. Marszałka Piłsudskiego 46, 81-378 Gdynia, Poland; paulina_lemke@wp.pl

Paulina Biskup is a PhD candidate in Earth sciences (oceanology) at the University of Gdańsk. Her main research interests include ecophysiology of marine algae, (pseudo)cryptic species, biogeography and biomonitoring. In her current research, PB focuses on the response of diatoms to salinity stress.

Magnus Boström School of Humanities, Education and Social Sciences, Örebro University, SE-701 82 Örebro, Sweden; magnus.bostrom@oru.se

Magnus Boström is professor in sociology and his research and teaching interests include politics, governance, participation, communication, organisation and responsibility in relation to transnational environmental and sustainability issues. Boström is also studying how various factors shape green consumerism and organised activism.

Marion Dreyer DIALOGIK, Lerchenstrasse 22, 70176 Stuttgart, Germany; dreyer@dialogik-expert.de

Marion Dreyer is deputy scientific director at DIALOGIK, which is a non-profit institute for communication and cooperation research. Her main fields of interest are risk governance and participation and cooperation processes in areas of societal controversy and conflict. In her ongoing research, MD looks at water issues in terms of precaution-based strategies to deal with anthropogenic micro-pollutants in the water cycle.

Sam Grönholm Åbo Akademi University, Turku, Finland; sgronhol@abo.fi

Sam Grönholm has an extensive academic working background and has been involved in numerous research projects, which have focused on sustainable development in the Baltic Sea region. He has also been employed as a project officer in pan-Baltic organisations, including the Union of the Baltic Cities and the Council of the Baltic Sea States. Currently, he works as a project coordinator at the Baltic University Programme Secretariat at the Uppsala Centre for Sustainable Development.

Björn Hassler School of Natural Sciences, Technology and Environmental Studies, Södertörn University, 14189 Huddinge, Sweden; bjorn.hassler@sh.se

Björn Hassler is associate professor in environmental science at Södertörn University and has extensive experience in multidisciplinary research on environmental governance in the Baltic Sea region. His main focus is on marine institutional structures in areas such as oil transportations, eutrophication and fisheries.

Anna Maria Jönsson School of Culture and Education, Södertörn University, 14189 Huddinge, Sweden; anna-maria.jonsson@sh.se

Anna Maria Jönsson is associate professor in media and communication studies at Södertörn University and has extensive experience in interdisciplinary research about environmental communication with a particular focus on journalism and risks as well as public participation in media and different governance processes.

Cecilia Lundberg Department of Biosciences, Environmental and Marine Biology, Åbo Akademi University, 20500 Åbo/Turku, Finland; cecilia.lundberg@abo.fi

Cecilia Lundberg is a PhD in marine biology with research interests in eutrophication and the Baltic Sea – from long-term changes of the water quality to management on an interdisciplinary level. CL is also involved in issues of higher education and is presently working part-time as a coordinator for the development of higher education at the Åbo Akademi University School of Business and Economics and part-time as a coordinator at the Centre of Lifelong Learning at Åbo Akademi University.

Piet Sellke DIALOGIK, Lerchenstrasse 22, 70176 Stuttgart, Germany; sellke@dialogik-expert.de

Piet Sellke has studied sociology and political science at the University of Stuttgart as well as sociology at the University of Oregon (USA). His main research interests are security studies, risk perception, risk governance and technology assessment. Currently, Piet Sellke is senior researcher and project manager with the DIALOGIK non-profit institute for communication and cooperation research.

Sara Söderström School of Natural Sciences, Technology and Environmental Studies, Södertörn University, 14189 Huddinge, Sweden; sara.soderstrom@sh.se

Sara Söderström is a PhD candidate in environmental science at Södertörn University. Her research interests are in regional environmental governance with a focus on the Baltic Sea, aligned with an emphasis on the ecosystem approach to management.

Aleksandra Zgrundo Department of Marine Ecosystem Functioning, Institute of Oceanography, University of Gdańsk, Al. Marszałka Piłsudskiego 46, 81-378 Gdynia, Poland; oceazg@ug.edu.pl

Aleksandra Zgrundo is an assistant professor at the University of Gdańsk (Poland). She is a marine biologist with an interest in integrating environmental sciences with political and social sciences. Her main fields of expertise include marine ecology and biodiversity and ecological assessment of marine waters.

Abbreviations

ACFA	Advisory Committee on Fisheries and Aquaculture
ACOM	The ICES Advisory Committee
AS	Alien species
AIS	HELCOMs Automatic Identification System
BEAT	HELCOM Biodiversity Assessment Tool
BFR	Brominated flame retardant
BNI	Baltic Nest Institute
BONUS+	Science for a Better Future of the Baltic Sea Region
BSAG	Baltic Sea Action Group
BSAP	Baltic Sea Action Plan
BSR	Baltic Sea region
BWMC	Ballast Water Management Convention
CAP	EU Common Agricultural Policy
CBD	Convention on Biological Diversity
CCB	Coalition Clean Baltic
CHASE	HELCOM Hazardous Substances Status Assessment Tool
CFP	EU Common Fisheries Policy
DG	Directorate-general
DG Environment	Directorate-general for the environment
DG MARE	Directorate-general for maritime affairs and fisheries
EAM	Ecosystem approach to management
EAFM	Ecosystem approach to fisheries management
EC	European Commission or European Community
EEA	European Environmental Agency
EEZ	Exclusive economic zone
ECHA	European Chemicals Agency
EMSA	European Maritime Safety Agency
EU	European Union
EUSBSR	EU Strategy for the Baltic Sea Region
GO	Governmental organisation
GOC	Global Ocean Commission

HEAT	HELCOM Eutrophication Assessment Tool
HELCOM	Helsinki Commission; Baltic Marine Environmental Protection Commission
IAS	Invasive alien species
IBSFC	International Baltic Sea Fisheries Commission
ICES	International Council for the Exploration of the Seas
IGO	International government organisation
ILO	International Labour Organisation
IMO	International Maritime Organisation
IPBES	Intergovernmental Platform on Biodiversity and Ecosystem Services
IPCC	Intergovernmental Panel on Climate Change
IPCP	International Panel on Chemical Pollution
LRF	Federation of Swedish Farmers
MARPOL	International Convention for the Prevention of Pollution from Ships
MPAs	Marine protected areas
MS	(EU) Member State
MSC	Marine Stewardship Council
MSFD	EU Marine Strategy Framework Directive
MSP	Marine spatial planning
MSY	Maximum sustainable yield
MTK	The Central Union of Agricultural Producers and Forest Owners
NATURA 2000	Habitats Directive and the Birds Directive
ND	EU Nitrates Directive
NFI	National Fisheries Institute
NGO	Non-governmental Organisation
NIP	National Implementation Program
OSPAR	Commission for protecting and conserving the North-East Atlantic and its resources
PA	Priority area
PCB	Polychlorinated biphenyl
PBDE	Polybrominated diphenyl ether
PFAS	Perfluoroalkylated substances
PFOA	Perfluorooctanoic acid
PFOS	Perfluorooctane sulphonate
PNS	Post-normal science
PSSA	Particularly sensitive sea area
RACs	Regional Advisory Councils, e.g. the Baltic Sea (BS) RAC
REACH	Registration, evaluation, authorisation and restriction of chemicals – the EU chemicals regulation
RISKGOV	Environmental Risk Governance of the Baltic Sea (Research Project within BONUS+)
SEPA	Swedish Environmental Protection Agency
STECF	Scientific, Technical, and Economic Committee for Fisheries

TAC	Total allowable catch
TGD	Technical guidance documents
UBC	Union of Baltic Cities
UNCLOS	United Nations Convention on the Law of the Sea
UWWTD	EU Urban Waste Water Treatment Directive
WFD	EU Water Framework Directive
WWF	World Wide Fund for Nature
WWT	Wastewater treatment

List of Figures

List of Tables

Chapter 1
Environmental Governance of the Baltic Sea: Identifying Key Challenges, Research Topics and Analytical Approaches

Michael Gilek, Mikael Karlsson, Sebastian Linke, and Katarzyna Smolarz

Abstract The Baltic Sea ecosystem is subject to a wide array of societal pressures and associated environmental risks (e.g. eutrophication, oil discharges, chemical pollution, overfishing and invasive alien species). Despite several years of substantial efforts by state and non-state actors, it is still highly unlikely that the regionally agreed environmental objectives of reaching "good environmental status" by 2021 in the HELCOM BSAP (Baltic Sea Action Plan) and by 2020 in the EU Marine Strategy Framework Directive (MSFD) will be met. This chapter identifies key research topics, as well as presents analytical perspectives for analysing the gap between knowledge and action in Baltic Sea environmental governance. It does so by outlining important trends and key challenges associated with Baltic Sea environmental governance, as well as by summarising the scope and results of individual chapters of this interdisciplinary volume. The analysis reveals the development of *increasingly complex governance arrangements* and the ongoing *implementation of the holistic Ecosystem Approach to Management*, as two general trends that together contribute to three key challenges associated with (1) *regional and cross-sectoral coordination and collaboration*, (2) *coping with complexity and uncertainty* in science-policy interactions and (3) *developing communication and knowledge sharing among stakeholder groups*. Furthermore, to facilitate analysis of environmental governance opportunities and obstacles both within and across

M. Gilek (✉) • M. Karlsson
School of Natural Sciences, Technology and Environmental Studies, Södertörn University, 14189 Huddinge, Sweden
e-mail: michael.gilek@sh.se; mikael.karlsson@2050.se

S. Linke
Department of Philosophy, Linguistics and Theory of Science, University of Gothenburg, Box 200, 405 30 Göteborg, Sweden
e-mail: sebastian.linke@gu.se

K. Smolarz
Department of Marine Ecosystem Functioning, Institute of Oceanography, University of Gdańsk, Al. Marszałka Piłsudskiego 46, 81-378 Gdynia, Poland
e-mail: oceksm@univ.gda.pl

M. Gilek et al. (eds.), *Environmental Governance of the Baltic Sea*,
MARE Publication Series 10, DOI 10.1007/978-3-319-27006-7_1

specific environmental issues, this chapter reviews the scientific literature to pinpoint key research issues and questions linked to the identified governance challenges.

Keywords Marine governance • Ecosystem approach to management • Institutional fit • Stakeholder participation • Science-policy interactions

1.1 Introduction

Governing marine environments is a highly complex and challenging enterprise (Gilek et al. 2015). This applies particularly to the heavily polluted and exploited, semi-enclosed and fragile Baltic Sea, situated in a densely populated region characterised by societal and ecological changes. This book aims for a better understanding of the complex arrangements of Baltic Sea environmental governance and gives proposals on how they could be developed for more sustainable outcomes. The book combines interdisciplinary investigations of the key environmental issues and risks in the area with in-depth analyses of problems, opportunities and barriers linked to governance structures and processes.

The Baltic Sea ecosystem is subject to a wide array of societal pressures such as hazardous chemicals, nutrients, oil discharges and invasive species, as well as exploitation of physical and biological resources such as fish (Ducrotoy and Elliott 2008; HELCOM 2010). For example, municipal wastewater, agricultural leakage and other sources have loaded the sea with phosphorus and nitrogen, which, together with intensive fishing and changing climate, have contributed to ecosystem regime shifts in some subbasins (Österblom et al. 2010) and a reduced capacity to deliver ecosystem goods and services to the people living in the nine coastal states (HELCOM 2010). Although most of these human pressures originate from activities in the Baltic Sea region (Fig. 1.1), significant contamination sources and other drivers of human-induced environmental change in the Baltic Sea also emanate from activities elsewhere and at larger scales, e.g. through long-range atmospheric transport of hazardous chemicals, introduction of invasive species through, e.g. shipping and global increases of anthropogenic greenhouse gases (e.g. Reckermann et al. 2012).

The coupled socioecological system associated with the Baltic Sea is today characterised by a dense multilevel web of governance structures (e.g. regulatory frameworks) and processes (such as science-policy interactions), which are linked to various forms of stakeholder participation and communication arrangements (e.g. Joas et al. 2008; Kern 2011). However, despite these thick layers of public and private governance arrangements, the Baltic Sea is still affected by serious environmental problems and risks due to various governance shortcomings (cf. HELCOM 2010). Furthermore, it is highly unlikely that the regionally agreed environmental objectives of reaching "good environmental status" by 2021 in the HELCOM BSAP

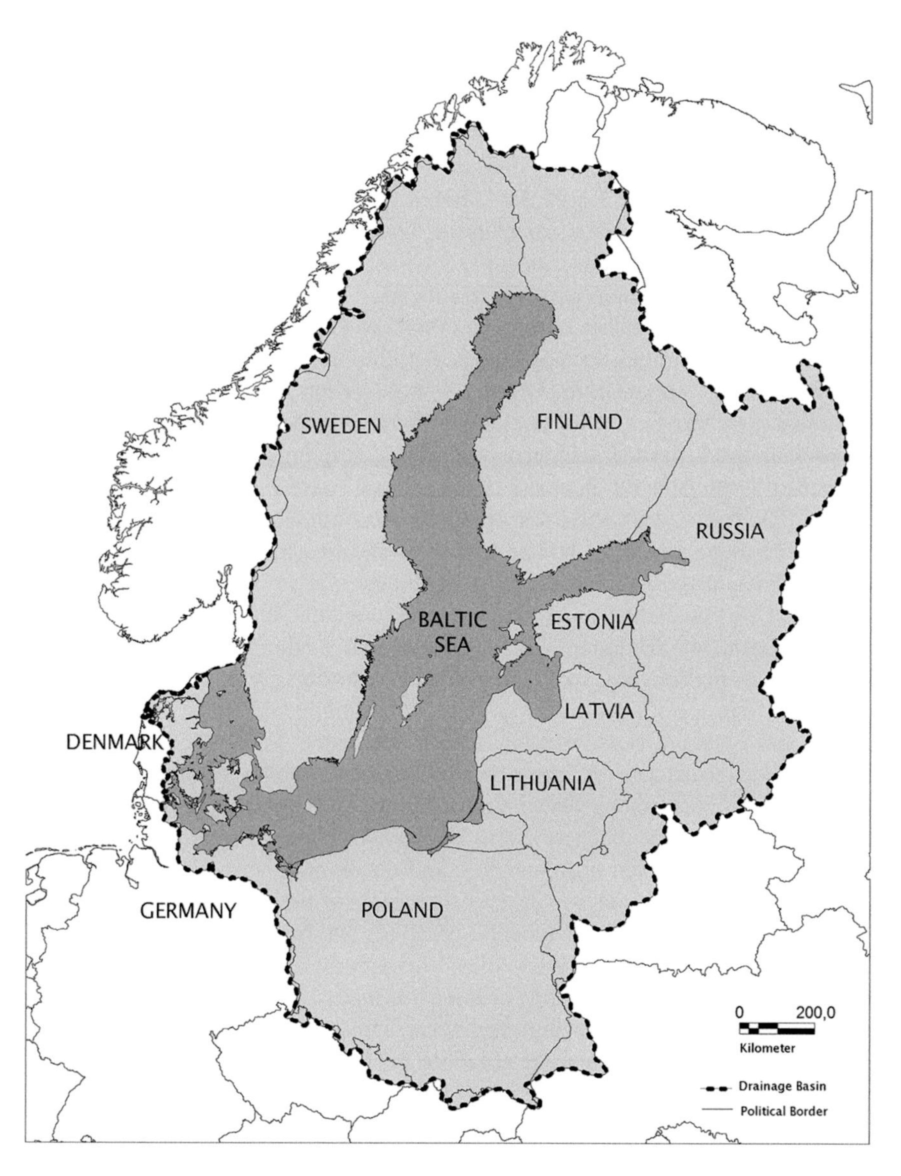

Fig. 1.1 The Baltic Sea region with its drainage basin and political borders (Modified from the GRID-Arendal Graphics Library, www.grida.no)

(Baltic Sea Action Plan) and by 2020 in the EU Marine Strategy Framework Directive (MSFD) will be met (Gilek and Kern 2011; Gilek et al. 2013; Kern 2011).

Consequently, this volume concentrates on the question of how key societal pressures and associated environmental risks (e.g. commercial fishing and the associated risks of overfishing, nutrient enrichment and eutrophication, shipping and oil discharges or invasive alien species) threatening the Baltic Sea environment are and could be governed. Our ultimate aim is to discuss pathways towards a more sustainable environmental governance of the Baltic Sea. Two general trends and associated challenges relating to environmental governance are of particular interest to the analysis.

First, linked to the complexity of human pressures and management responses in the Baltic Sea region, significant differences have evolved in the *governance frameworks* of various environmental issues, over time and between problem areas. The chapters of this book describe and analyse the evolution of this complex web of Baltic Sea environmental governance structures, through comparative investigations of in-depth case studies of five important problems and risks: eutrophication, overfishing, invasive alien species, chemical pollution and oil discharges.

Second, the *Ecosystem Approach to Management* (EAM) is today widely acclaimed in science and policy circles worldwide as a means to integrate measures in order to reach desired socioeconomic and environmental objectives, thereby facilitating sustainable development of marine and coastal areas (e.g. Backer et al. 2010; CBD 1998; Curtin and Prellezo 2010). According to this holistic approach, sustainable management of human activities and pressures should be based on the specific sensitivity and complexity of the ecosystem in focus, as well as on integration of cumulative pressures, e.g. over various sources of pollution and resource extraction (Hammer 2015; McLeod and Leslie 2009). Central to the concept is that management should be based on all forms of relevant knowledge and experience (e.g. scientific, local, actor-based knowledge), as well as on stakeholder participation, precaution and adaptability (cf. Hammer et al. 2011). However, since EAM is a very broad concept, views on what it exactly implies and how it should be implemented varies among and within different stakeholder groups, as well as among various groups of experts and researchers contributing to science-based advice. This multifaceted understanding and the framing and implementation of EAM is described and analysed in several chapters of the book with respect to the governance of particular environmental problems and risks, as well as in terms of challenges for processes of science-policy interactions, stakeholder communication and participation. In the concluding chapter, this discussion on problems and opportunities associated with achieving integration, across, for example, levels, sectors, interests and knowledge claims, is expanded to also include an attempt to identify broader pathways, as well as concrete institutional reforms and strategies that potentially could strengthen EAM implementation and outcomes.

Despite (and to some extent as a consequence of) these trends, i.e. the development of complex governance arrangements and the adoption of EAM, a number of key challenges remain as important obstacles for achieving sustainable governance of the Baltic Sea (Gilek et al. 2011), as well as marine ecosystems elsewhere (e.g. Gilek and Kern 2015). Of particular interest to the aims of this volume are three challenges that relate to *multilevel*, *knowledge-based* and *inclusive* environmental governance of the Baltic Sea. The first challenge concerns *difficulties to establish adequate regional cross-sectoral collaboration* among Baltic Sea policy actors due to existing institutional structures and procedures, power relations, cultures and varying policy styles in the different countries of the region. The second challenge concerns the *uncertainties, ambiguities and complexities* involved in perceiving, understanding and assessing different types of risks and problems, as well as *how these risk perceptions and science-based assessments interact with environmental management*. The third challenge concerns the *difficulties in developing communication, exchange of value perspectives and knowledge sharing* among key stakeholder groups based on participation, transparency and trust.

In this chapter we first introduce environmental governance and the key governance challenges identified and addressed in the book, as well as important research topics associated with these challenges. Second, we outline the general analytical and methodological approaches on which the research presented in this book is based. Finally, we summarise the book's structure and highlight key topics addressed in the individual chapters.

1.2 Key Environmental Governance Challenges and Related Research Topics

Over the last few decades, the term 'governance' and the specific topic of this book – 'environmental governance' – have become prevalent in the social and environmental science literature (e.g. Söderström et al. 2015a). The concept of governance, first established in public administration and taken up in political science, is used to depict a shift in responsibility from state to non-state actors (e.g. private or voluntary sectors) that affects structures and processes for collective action and decision-making (Stoker 1998). It emphasises social and political steering and acting in polycentric networks on different levels – local, regional, national, European and global (e.g. Delmas and Young 2009; Rosenau 2003; Wagner 2005). However, there is a great variation in how governance is defined and used. In other words, it refers more to a perspective than a coherent theory (e.g. Pierre and Peters 2000; Rhodes 1996). Our definition in this volume includes both structures – such as policy contexts, existing power relations among key actors, regulatory frameworks and organisational forms of decision-making, reflexivity and participation – and processes. Processes comprise aspects such as the evolution of organisations and interactions between, for instance, science and policy, as well as communication and interaction among policy-makers, scientists and other stakeholders. Processes also include the development of strategies, framings, communication and learning.

In many respects, EAM shares with environmental governance an interest in similar core topics, e.g. multilevel and multi-sector institutional interactions, knowledge integration and stakeholder arrangements and partnerships (Söderström et al. 2015a). In addition, it offers several additional focal areas and assumptions of value to the comprehensive analysis of environmental governance aimed for in this book. To begin with, there is a fundamental ecosystem-based focus in EAM that assumes that sustainable management of human activities and pressures can only be achieved if it is based on the sensitivity and complexity of the ecosystem in focus. In line with this, a central idea in EAM is that management needs to be based on congruence between institutions and ecosystems, as well as between institutions and environmental problems (institutional fit) (e.g. Folke et al. 2007). Furthermore, EAM aims to address environmental issues and their management in a holistic and integrated manner (e.g. McLeod and Leslie 2009), implying that the concept promotes an approach that analyses multiple objectives (e.g. socio-economic developments and environmental status), as well as multiple sectoral interests (e.g. fisheries, maritime transports, tourism, etc.). Finally, linked to discussions on the need for adaptive co-management (e.g. Armitage et al. 2007), EAM offers approaches to analyse prerequisites and implications of adaptation, collaboration and learning linked to multilevel stakeholder arrangements and science-policy interfaces.

Research has also shown that the governance of environmental problems and risks[1] in, for example, marine areas poses specific challenges and problems in that they (1) usually exhibit extremely complex multilevel interactions between risk-causing human activities and societal responses to these (Gilek and Kern 2015); (2) usually are associated with a striking scientific uncertainty (Udovyk and Gilek 2013); and (3) are characterised by social complexity which requires substantial, not seldomly contested, debate on what is at stake, what choices to make and which values are being assigned to different components of the ecosystem and to various strategies (Lidskog et al. 2009; van Asselt and Renn 2011). Hence, based on insights on governance in general and on EAM and environmental issues in particular (e.g. Lidskog et al. 2009), this book focuses on three key governance dimensions and challenges: *multilevel and multi-sectoral governance structures*, *assessment-management processes and interactions and stakeholder participation and communication*, as discussed below.

1.2.1 Multilevel and Multi-sectoral Governance Structures

Environmental governance in general and marine governance in particular are characterised and challenged by complex multilevel and multi-sectoral interactions (cf. Gilek and Kern 2015; Lidskog et al. 2009). The Baltic Sea environmental governance system is, for example, made up of structures of national, international, European and transnational governance and can be perceived of as the outcome of continuous disparate processes over time, rather than being part of an intentionally designed governance arrangement (Andonova and Mitchell 2010). Furthermore, although marine environmental governance has traditionally focused on particular

[1] It can be argued that environmental issues, even if they already manifested themselves as negative environmental impacts, are associated with significant uncertainties about the type and extent of impacts, probabilities for future impacts, effectiveness of management responses, etc. Therefore, we argue in this volume that the risk governance concept, which acknowledges the central role of ignorance, uncertainty and ambiguity in decision-making on risks (Stirling 2007; Renn 2008) provides a suitable analytical perspective.

sources of contamination (e.g. hazardous chemicals) and the use of natural resources (e.g. commercial fisheries), there are now strong policy ambitions to achieve multi-sector integration through approaches such as EAM, integrated ecosystem assessments (IEA) and marine spatial planning (MSP) (Douvere 2008; Karlsson et al. 2011; Linke et al. 2014; Udovyk and Gilek 2014; Walther and Möllmann 2014). Hence, understanding the processes and outcomes of Baltic Sea environmental governance requires that several multilevel and multi-sector interactions and associated challenges are simultaneously considered.

First, at the national level environmental governance may vary considerably among the states surrounding the Baltic Sea, which complicates international collaboration (e.g. Gilek et al. 2013). In the Baltic Sea region, we find countries such as Sweden that have gained a reputation as environmental pioneers since the 1970s and countries such as Poland and the three Baltic states that started to develop their environmental policy with a background of having been centrally planned economies.

Second, beyond the national level, Baltic Sea environmental governance is affected not only by global and EU agreements (such as the MARPOL Convention, EU regulations, directives and policies) but also by the regional international Helsinki Convention and its Baltic Sea Action Plan (BSAP) which came into effect from 2007 (HELCOM 2007). Even though this regionalisation of marine governance at the level of the entire Baltic Sea has the potential to improve multilevel coordination and cooperation by, for example, distinguishing between measures that can be implemented at international, EU and national levels, the successful harmonisation and coordination of actions still remain to be done (Gilek and Kern 2011). It is also possible that differences in, for example, path dependency, power relations and knowledge base will lead to differences in efficiency and outcomes of environmental governance at the regional level in various sectors (e.g. fisheries, shipping, agriculture, etc., cf. Linke et al. 2014).

Third, the Europeanisation of the Baltic Sea has developed quickly. This is most prominent in the area of fisheries as witnessed by the dominance of the EU Common Fisheries Policy (CFP) (Linke et al. 2014), but includes a proliferating body of EU legislation affecting various aspects of the marine environment under the guidance of EAM (cf. Raakjær and Tatenhove 2014). However, there is a division between EU policies that aim primarily at achieving good environmental status, such as the Marine Strategy Framework Directive (MSFD), and those aiming to regulate pollutants (e.g. the REACH chemicals regulation) and the use of natural resources, such as fish (e.g. EU CFP). This points to the need to achieve integration of various EU policies, since different policy objectives may lead to contradictions and conflicts between, for example, fisheries and marine nature conservation (De Santo 2015).

All these aspects are explored in this book. A particular focus is, however, placed on the macro-regional Baltic Sea level, because the most severe environmental issues in the Baltic Sea such as eutrophication and chemical pollution affect the Baltic ecosystem at large spatial scales that transgress national borders (HELCOM 2010). Moreover, national and local management measures are in practice often based on decisions at supranational levels. It has been argued that analyses of regional and macro-regional environmental governance (e.g. at the scale of the entire Baltic Sea) are less prevalent in the scientific literature on environmental governance than both at the local and global levels (e.g. Balsiger and Debarbieux 2011; Gilek and Kern 2015), which underlines the need for the regional perspective explored in this volume.

1.2.2 Assessment: Management Processes and Interactions

The interactions between the primarily science-based assessment sphere (i.e. generation of knowledge on environmental status, pressures, risks and problems) and the management sphere (i.e. decisions on and implementation of actions) have been described as key processes in environmental governance (Renn et al. 2011; Rice 2005). Science has since long been seen as the primary provider of knowledge and advice to guide environmental policy-making, especially in the case of managing environmental risks stemming from industrial technologies and pollutants (Karlsson et al. 2011). This has also been the case in the Baltic Sea region, both nationally and in relation to the activities of international organisations such as HELCOM and ICES (Udovyk and Gilek 2013).

However, interactions over science-policy interfaces (e.g. connected with the evaluation of what constitutes good environmental status and unacceptable levels of risk) are usually complicated by severe challenges connected with complexity, ignorance, uncertainty and ambiguity (Renn 2008; Stirling 2007), which frequently result in controversy in both society and science on appropriate risk assessment and management. It has been argued that scientific uncertainties and stakeholder disagreements and conflicts are particularly problematic for marine environmental governance when implementing holistic management approaches such as EAM and MSP (Linke et al. 2014; Rice 2005; Wilson 2009). Observations of impaired public trust in science and recognition of other legitimate knowledge providers, such as practitioners, stakeholders and experts based elsewhere than in traditional research organisations, have also been linked to cases of severe scientific uncertainty (Irwin and Michael 2003), in combination with a common politicisation of science (e.g. Eriksson et al. 2010; Weingart 1999). In response, Stirling (2007)[2] has argued that different types of environmental issues characterised by uncertainty and ambiguity require an expansion of traditional strategies in science and policy, to include precautionary and participative approaches.

As a consequence, the relationship between science and policy is changing on both a theoretical and practical level, particularly with regard to complex environmental issues such as marine governance. It is, however, despite a long and

[2] Stirling (2007) differentiates between four types of scientific incertitude: risk (quantitative data and knowledge exist), uncertainty (qualitative understanding of outcome, but not probabilities), ambiguity (poor knowledge about potential outcome) and ignorance; see Linke et al. (2016) for further explanation.

strong tradition of scientific exploration of the Baltic Sea, still unclear if and how these changes will affect environmental governance issues in the Baltic Sea. Key questions in this context are with regard to if and how strategies for coping with the fundamental problems of different kinds of uncertainty have evolved for particular issues and sectors. This book will investigate these questions in-depth.

1.2.3 Stakeholder Participation and Communication

Various actors (e.g. policy-makers, social scientists, civil society organisations, etc.) generally agree that for societies to be able to manage and govern large-scale environmental risks, there is a need for transnational communication and multi-stakeholder participation, as well as for increased involvement of citizens through various processes of deliberation. For example, several scholars have argued for the need to facilitate stakeholder inclusion and deliberation in the governance of the marine environment and natural resources such as fish (cf. Jentoft and Chuenpagdee 2015; Mackinson et al. 2011). To facilitate stakeholder inclusion and participation, several new institutional arrangements have also been discussed, such as 'joint environmental policy-making' (Mol et al. 2000), 'multi-stakeholder dialogue' (Bendell 2000) and 'partnership' (Glasbergen et al. 2007). In the governance of regional seas such as the Baltic Sea, collaboration fostering initiatives by non-governmental and subnational organisations (Kern and Löffelsend 2004), as well as transnational stakeholder networks, have also been found to be influential in many environmental governance contexts (Kern and Bulkeley 2009). Adding to this complexity, the institutions for Baltic Sea environmental governance have developed rather rapidly over the last few years in the form of venues for stakeholder participation such as the Regional Advisory Councils (RACs) in EU fisheries management (e.g. Linke et al. 2011) and stakeholder forums organised by HELCOM (e.g. Hassler et al. 2013). A core question is, however, to what extent these complex stakeholder arrangements open up collaboration and learning as opposed to impede possibilities to, for example, bridge sectoral interests.

Previous research on environmental governance has revealed several benefits of inclusive, communicative and participatory approaches (e.g. Lafferty and Meadowcroft 1996; Lovan et al. 2004), but also situations when participation may not be successful (e.g. Boström 2006). The advantages of more inclusive governance approaches relate to normative and instrumental reasons. Broad inclusion can be seen as normatively (intrinsically) 'good' because the idea of inclusiveness responds to democratic ideals around socially just representation. Citizens that are potentially affected by, for example, environmental pollution should be given access to data and processes and provided with opportunities to voice their concerns in communicative and even judicial forums, a principle established, for example, by the Aarhus convention (UN ECE 1998). The academic literature also discusses instrumental reasons for inclusiveness (e.g. Boström 2006; Jönsson et al. 2016). For example, it has been argued that inclusiveness generates new and more socially robust knowledge, stimulates mutual learning and ultimately facilitates capacity building in environmental governance.

Still, despite a basic descriptive understanding of the complex stakeholder arrangements and their recent developments in the Baltic Sea region (e.g. Hassler et al. 2013; Kern and Bulkeley 2009), there is a need for more in-depth critical analyses of framings, processes and outcomes linked to stakeholder participation in Baltic Sea environmental governance. Similarly, knowledge on environmental communication and framing is rather undeveloped in the Baltic Sea region, although some previous studies have addressed, for example, media framing (Jönsson 2011) and stakeholder participation in fisheries management (Linke et al. 2011). Clearly, stakeholders' perceptions, engagement and participation can all be influenced by how the Baltic Sea environment and its problems are communicated and framed in the public discourse (cf. Cox 2006). In particular, in line with this book's ambition to understand environmental governance structures and processes at the macro-regional level of the Baltic Sea, there is a need to better understand the extent to which there are supranational communication arenas in the Baltic Sea region. These questions and perspectives relating to stakeholder participation and communication are all covered in the book and applied to experiences of environmental governance in the Baltic Sea region.

1.3 Outline of Analytical and Methodological Approaches

The empirical work presented in the chapters of this book was gathered as part of the interdisciplinary RISKGOV project[3] which was based on a common analytical and methodological framework. Furthermore, empirical and analytical insights from the 'follow-up' COOP project[4] were used to update and expand several case studies such as the one on eutrophication, as well as to develop cross-case comparisons and ideas for improvements.

The analytical framework aimed to ensure possibilities for cross-case comparisons by specifying focused governance dimensions in line with the arguments presented in Sect. 1.2 above, defining main research questions and providing the methodological requirements for interviews and document studies. These analytical and methodological specifications are outlined below. While reading this book, however, it is important to note that the authors of the individual chapters have been asked to focus on particularly important and interesting aspects in their respective cases. This means that the main aim of this volume is to explore challenging aspects

[3] RISKGOV (Environmental Risk Governance of the Baltic Sea) was funded by the BONUS+ programme and the Foundation for Baltic and East European Studies (2009–2015). www.sh.se/riskgov.

[4] COOP (Cooperating for Sustainable Regional Marine Governance) was funded by the Foundation for Baltic and East European Studies (2012–2015).

associated with the different cases of environmental governance in the Baltic Sea, rather than to strive for full-fledged cross-case comparisons.[5]

To start with, five key environmental issues and risks from the Baltic Sea were identified for in-depth case studies: eutrophication, overfishing, invasive alien species, chemical pollution and oil discharges linked to marine transports. These issues have all been shown to be major, large-scale environmental problems in the Baltic Sea and are prioritised in national, regional (e.g. BSAP) as well as European (e.g. MSFD) marine regulatory frameworks (cf. Söderström et al. 2015b). Moreover, these cases represent a variety of types of environmental problems in terms of, for example, complexity of causes, scientific uncertainty and sociopolitical controversy (as will be described and analysed in the chapters of this volume).

The insights on governance in general and on environmental issues in particular described in Sect. 1.2 were the motivation behind choosing the three governance dimensions of primary design and analytic importance in the project: *multilevel and multi-sectoral governance structures*, *assessment-management processes and interactions and stakeholder participation and communication* (Fig. 1.2). Hence, the aim has been to study both the horizontal axis of risk governance focusing on a plurality of actors and norms and the vertical axis focusing on the connections and interactions between different scales in space and time (e.g. Lyall and Tait 2004). This means that although the main focus has been on the regional (i.e. transnational) Baltic Sea scale, interlinkages with other important levels such as nation states, the EU and global collaboration have been included to facilitate a comprehensive understanding of environmental governance of the Baltic Sea. In other words, the focus is on Baltic Sea *regional environmental governance*, but without losing sight of the relevance of other policy levels. Key research issues and governance challenges associated with the focused governance dimensions are further specified in Table 1.1.

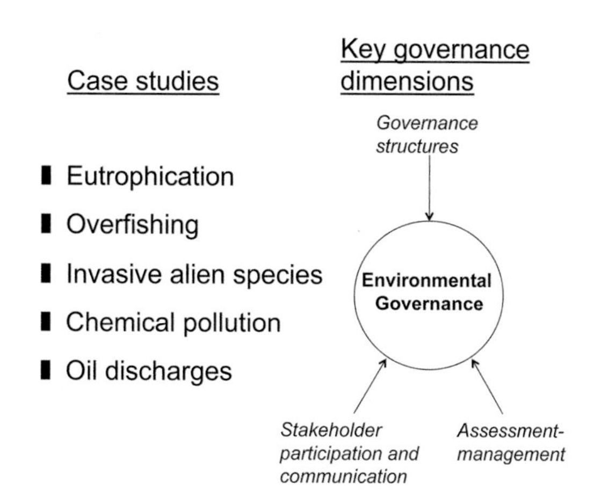

Fig. 1.2 Outline of case studies of environmental governance and the key governance dimensions of Baltic Sea environmental governance analysed in this book

[5] For other results from RISKGOV, please see project reports published on www.sh.se/riskgov.

Table 1.1 Key research issues and environmental governance challenges linked to the studied dimensions of Baltic Sea environmental governance

Governance dimensions	Identified main governance challenges	Key research issues
Multi-level and multi-sectoral governance structures	Difficulties to establish adequate regional cross-sectoral collaboration	Governmental organisations and networks
		Non-governmental organisations and networks
		Key policy documents and regulatory frameworks
		Multi-level and multi-sector interactions
Assessment – management processes and interactions	Difficulties to cope with uncertainties, ambiguities and complexities in environmental governance	Organisation of science-policy interactions; role of science-based advice
		Coping with uncertainty and disagreements
Stakeholder participation and communication	Difficulties in developing communication and knowledge sharing among key stakeholder groups	Problem and media framing
		Arrangements for stakeholder/public involvement and communication

In terms of empirical material, the five case studies of Baltic Sea environmental governance (Fig. 1.2) were based on a combination of several types of data sources acquired during 2009–2014: text analysis of key documents, interviews of key informants and roundtables. The case study work was organised in three consecutive steps guided by the analytical framework.

First, linked to each case study (cf. Fig. 1.2), a cross-disciplinary team of researchers (e.g. based in environmental science, ecotoxicology, environmental sociology, political science or media and communication studies) conducted a review of secondary material (existing empirical literature on each focused Baltic Sea environmental problem), as well as of primary sources such as documents and data bases on governance structures, problem assessment and stakeholder communication processes.

Second, each case study research team conducted interviews – approximately 15 per case – with key experts representing governmental and non-governmental organisations, comprising important parts of the governance structure of each issue area. The interviews were semi-structured (approx. 1–2 h) and developed in close collaboration with the different case study teams to facilitate comparability.

Third, to facilitate cross-case comparisons, three joint thematic round-table discussions (on regional governance structures, scientific knowledge and uncertainty and stakeholder participation and communication) were arranged in collaboration among all case study teams to get further input on similarities and differences among cases from experts, decision-makers and other stakeholders.

1.4 Structure and Content of the Book

Following this introduction (Chap. 1), the book is divided into two parts. In the first part, five in-depth interdisciplinary case studies of environmental governance associated with large-scale environmental problems and risks in the Baltic Sea region (i.e. eutrophication, overfishing, invasive alien species, chemical pollution and oil discharges) are presented and analysed. In the second part of the book, key challenges and possible avenues for improvements are identified and analysed across the covered environmental issues, based on the three governance dimensions identified (Fig. 1.2; Table 1.1). Particular emphasis is placed on challenges for EAM implementation linked to multilevel and multi-sector environmental governance, science-policy interfaces, as well as stakeholder communication and participation.

The first part of the book starts with the case of eutrophication, one of the most serious environmental problems in the Baltic Sea. *Karlsson and co-workers* (Chap. 2) describe the complex problem and the governance structures and processes in place. The case shows how science-policy interactions have so far worked comparatively well, resulting in an ongoing implementation of EAM, although fundamental societal change is still needed in order to reach agreed objectives. In Chap. 3, *Sellke and colleagues* analyse the fisheries case, where scientific uncertainty, a multitude of actors representing contradictory interests and the tensions between top-down EU and bottom-up regional policies may paralyse decision-making. By pointing out the most pressing issues, the authors aim to provide input that may contribute to improving fisheries governance. In Chap. 4, *Smolarz and co-workers* take on the case of invasive alien species and describe the striking uncertainty on the one hand and the low interest among policy-makers and stakeholders on the other. In elaborating on a governance framework, including voluntary measures and improved coordination of public policies at various levels, the authors give attention to a severe problem that cannot continue to be neglected. Uncertainty is a striking component also in the following case of hazardous chemicals (Chap. 5). *Karlsson and Gilek* zoom in on the governance of three specific organohalogens that have caused severe problems and risks in the Baltic Sea. The authors compare measures taken over time by EU and HELCOM, respectively, and analyse what those experiences might mean for improving public governance in the future. Finally, *Hassler* (Chap. 6) identifies the primary drivers behind accidental oil spills and intentional oil discharges into the Baltic Sea. The author makes a case for global conventions and coordinated Port State Control in the former case and development of changed incentives for operators in the latter case, e.g. by institutionalising no-special-fee systems for waste management in ports.

The second part of the book is structured according to the three governance dimensions (governance structures, assessment-management processes and interactions and stakeholder participation and communication, see Fig. 1.2). Each chapter discusses and compares certain characteristics of the five specific cases presented in the first part of the book. In Chap. 7, *Boström and colleagues* describe the evolution of governance structures over time up to the present-day ambitions of implementing

EAM and investigate if the present institutional and regulatory set-up supports EAM implementation. The authors apply the concept of *reflexive governance* in order to analyse various governance modes and elaborate on possible pathways to make Baltic Sea environmental governance more sustainable. Next, *Linke and colleagues* (Chap. 8) compare the science-policy interfaces linked to the five cases, with a focus on organisational structures and management of uncertainties and stakeholder disagreements. On that basis, possible routes for improving interaction between science-based advice and environmental management are discussed, in particular regarding implementation of the EAM. Finally, *Jönsson and co-workers* (Chap. 9) analyse how risks and problems are framed in the five cases and what role communication plays in the governance context with respect to institutions and procedures. The results point to the importance of widening the stakeholder concept and acknowledging the importance of citizen and public communication in practice.

Following the first and second part of the book, *Gilek and Karlsson* (Chap. 10) draw from the conclusions and recommendations of previous chapters to identify root problems and possible pathways for improving environmental governance in the Baltic Sea.

Acknowledgements The research leading to these results was funded by the Foundation for Baltic and East European Studies and the European Community's Seventh Framework Programme (2007–2013) under grant agreement n° 217246 made with the joint Baltic Sea research and development programme BONUS, as well as by the German Federal Ministry of Education and Research (BMBF), the Swedish Environmental Protection Agency, the Swedish Research Council FORMAS, the Polish Ministry of Science and Higher Education and the Academy of Finland. We wish to thank these institutions for enabling this research. Two peer reviewers are also thanked for valuable comments on an earlier version of the chapter.

References

Andonova LB, Mitchell RB (2010) The rescaling of global environmental politics. Ann Rev Environ Resour 35:255–282

Armitage DR, Berkes F, Doubleday N (eds) (2007) Adaptive co-management: collaboration, learning, and multi-level governance. UBC Press, Vancouver

Backer H, Leppänen JM, Brusendorff AC, Forsius K, Stankiewicz M, Mehtonen J, Pyhälä M, Laamanen M, Paulomäki H, Vlasov N, Haaranen T (2010) HELCOM Baltic Sea action plan – a regional programme of measures for the marine environment based on the ecosystem approach. Mar Pollut Bull 60:642–649

Balsiger J, Debarbieux B (2011) Major challenges in regional environmental governance research and practice. Procedia Soc Behav Sci 14:1–8

Bendell J (ed) (2000) Terms for endearment: business, NGOs and sustainable development. Greanleaf Publishing, Sheffield

Boström M (2006) Regulatory credibility and authority through inclusiveness. Standardization organizations in cases of eco-labelling. Organization 13(3):345–367

CBD (Convention on Biodiversity) (1998) Report of the workshop on the ecosystem approach. Lilongwe. 26–28 Jan 1998. UNEP/COP/4/Inf.9

Cox R (2006) Environmental communication and the public sphere. SAGE Publications Ltd, London

Curtin R, Prellezo R (2010) Understanding marine ecosystem based management: a literature review. Mar Policy 34:821–830

De Santo E (2015) The marine strategy framework directive as a catalyst for maritime spatial planning: internal dimensions and institutional tensions. In: Gilek M, Kern K (eds) Governing Europe's marine environment. Europeanization of regional seas or regionalization of EU policies? Ashgate Publishing, Farnham

Delmas MA, Young OR (eds) (2009) Governance for the environment. New perspectives. Cambridge University Press, Cambridge

Douvere F (2008) The importance of marine spatial planning in advancing ecosystem-based sea use management. Mar Policy 32(5):762–771

Ducrotoy JP, Elliott M (2008) The science and management of the North Sea and the Baltic Sea: natural history, present threats and future challenges. Mar Pollut Bull 57:8–21

Eriksson J, Karlsson M, Reuter M (2010) Technocracy, politicization, and non-involvement: politics of expertise in the European regulation of chemicals. Rev Policy Res 27:167–185

Folke C, Pritchard L, Berkes F, Colding J, Svedin U (2007) The problem of fit between ecosystems and institutions: ten years later. Ecol Soc 12(1):30, http://www.ecologyandsociety.org/vol12/iss1/art30/

Gilek M, Kern K (2011) Regional issues: environment. In: Henningsen B, Etzold T (eds) Political state of the region report 2011. Baltic Development Forum, Copenhagen, pp 68–71

Gilek M, Kern K (eds) (2015) Governing Europe's marine environment. Europeanization of regional seas or regionalization of EU policies? Ashgate Publishing, Farnham

Gilek M, Hassler B, Jönsson AM, Karlsson M (eds) (2011) Coping with complexity in Baltic Sea risk governance. Ambio 40(2):109–110

Gilek M, Hassler B, Engkvist F, Kern K (2013) The HELCOM Baltic Sea action plan: challenges of implementing an innovative ecosystem approach. In: Henningsen B, Etzold T, Pohl AL (eds) Political state of the region report 2013 – trends and directions in the Baltic Sea region. Baltic Development Forum, Copenhagen, pp 58–61

Gilek M, Hassler B, Jentoft S (2015) Marine environmental governance in Europe: problems and opportunities. In: Gilek M, Kern K (eds) Governing Europe's marine environment. Europeanization of regional seas or regionalization of EU policies? Ashgate Publishing, Farnham

Glasbergen P, Biermann F, Mol APJ (eds) (2007) Partnerships, governance and sustainable development. Reflections on theory and practice. Edward Elgar, Cheltenham

Hammer M (2015) The ecosystem management approach. Implications for marine governance. In: Gilek M, Kern K (eds) Governing Europe's marine environment. Europeanization of regional seas or regionalization of EU policies? Ashgate Publishing, Farnham

Hammer M, Balfors B, Mörtberg U, Petersson M, Quin A (2011) Governance of water resources in the phase of change: a case study of the implementation of the EU Water Directive in Sweden. Ambio 40(2):210–220

Hassler B, Boström M, Grönholm S (2013) Towards an ecosystem approach to management in regional marine governance? The Baltic Sea context. J Environ Policy Plan 15(2):225–245

HELCOM (2007) HELCOM Baltic Sea action plan. Available from: http://helcom.fi/Documents/Baltic%20sea%20action%20plan/BSAP_Final.pdf

HELCOM (2010) Ecosystem health of the Baltic Sea 2003–2007: HELCOM initial holistic assessment. Baltic Sea Environmental Proceedings No. 122
Irwin A, Michael M (2003) Science, social theory and public knowledge. Open University Press, Maidenhead
Jentoft S, Chuenpagdee R (2015) The 'new' marine governance: assessing governability. In: Gilek M, Kern K (eds) Governing Europe's marine environment. Europeanization of regional seas or regionalization of EU policies? Ashgate Publishing, Farnham
Joas M, Jahn D, Kern K (eds) (2008) Governing a common sea. Environmental policies in the Baltic Sea region. Earthscan, London
Jönsson AM (2011) Framing environmental risks in the Baltic Sea – a news media analysis. Ambio 40:121–132
Jönsson AM, Boström M, Dreyer M, Söderström S (2016) Risk communication and the role of the public: towards inclusive environmental governance of the Baltic Sea? In: Gilek M (ed) Environmental governance of the Baltic Sea. Springer, Dordrecht
Karlsson M, Gilek M, Udovyk O (2011) Governance of complex socio-environmental risks: the case of hazardous chemicals in the Baltic Sea. Ambio 40(2):144–157
Kern K (2011) Governance for sustainable development in the Baltic Sea region. J Balt Stud 42(1):67–81
Kern K, Bulkely H (2009) Cities, Europeanization and multi-level governance: governing climate change through transnational municipal networks. J Common Mark Stud 47(1):309–332
Kern K, Löffelsend T (2004) Sustainable development in the Baltic Sea region. Governance beyond the nation state. Local Environ 9(5):451–467
Lafferty WM, Meadowcroft J (eds) (1996) Democracy and the environment: problems and prospects. Edward Elgar, Cheltenham
Lidskog R, Soneryd L, Uggla Y (2009) Transboundary risk governance. Earthscan, London
Linke S, Dreyer M, Sellke P (2011) The regional advisory councils. What is their potential to incorporate stakeholder knowledge into fisheries governance? Ambio 40(2):133–144
Linke S, Gilek M, Karlsson M, Udovyk O (2014) Unravelling science-policy interactions in environmental risk governance of the Baltic Sea: comparing fisheries and eutrophication. J Risk Res 17(4):505–523
Linke S, Gilek M, Karlsson M (2016) Science-policy interfaces in Baltic Sea environmental governance: towards regional cooperation and management of uncertainty? In: Gilek M et al (eds) Environmental governance of the Baltic Sea. Springer, Dordrecht
Lovan R, Murray M, Shaffer R (eds) (2004) Participatory governance: planning, conflict mediation and public decision-making in civil society. Ashgate, Aldershot
Lyall C, Tait J (2004) Shifting policy debates and the implications for governance. In: Lyall C, Tait J (eds) New modes of governance. Developing an integrated policy approach to science, technology, risk and the environment. Ashgate, Aldershot, pp 3–17
Mackinson S, Wilson DC, Galiay P, Deas B (2011) Engaging stakeholders in fisheries and marine research. Mar Policy 35:18–24
McLeod K, Leslie H (eds) (2009) Ecosystem-based management for the oceans. Island Press, Washington, DC
Mol APJ, Lauber V, Liefferink D (eds) (2000) The voluntary approach to environmental policy: joint environmental policy-making in Europe. Oxford University Press, Oxford
Österblom H, Gårdmark A, Bergström L, Müller-Karulis B, Folke C, Lindegren M, Casisni M, Olsson P, Diekmann R, Blenckner T, Humborg C, Möllmann C (2010) Making the ecosystem approach operational. Can regime shifts in ecological- and governance systems facilitate the transitions? Mar Policy 34:1290–1299
Pierre J, Peters BG (2000) Governance, politics and the state. Macmillan Press, Basingstove
Raakjær J, van Tatenhove J (2014) Marine governance of European seas: introduction. Mar Policy 50:323–324
Reckermann M, Brander K, MacKenzie BR, Omstedt A (eds) (2012) Climate impacts on the Baltic Sea: from science to policy. Springer, Berlin
Renn O (2008) Risk governance: coping with uncertainty in a complex world. Earthscan, London

Renn O, Klinke A, van Asselt M (2011) Coping with complexity, uncertainty and ambiguity in risk governance: a synthesis. Ambio 40(2):231–246
Rhodes RAW (1996) The new governance: governing without government. Polit Stud 44:652–667
Rice JC (2005) Implementation of the ecosystem approach to fisheries management – asynchronous co-evolution at the interface between science and policy. Mar Ecol Prog Ser 300:265–270
Rosenau JN (2003) Distant proximities. Dynamics beyond globalization. Princeton University Press, Princeton
Söderström S, Kern K, Boström M, Gilek M (2015a) Environmental governance and ecosystem management: avenues for synergies between two approaches. Interdisc Environ Rev (2016)
Söderström S, Kern K, Hassler B (2015b) Marine governance in the Baltic Sea: current trends of Europeanization and regionalization. In: Gilek M, Kern K (eds) Governing Europe's marine environment. Europeanization of regional seas or regionalization of EU policies? Ashgate Publishing, Farnham
Stirling A (2007) Risk, precaution and science: towards a more constructive policy debate. EMBO Rep 8:309–315
Stoker G (1998) Governance as theory: five propositions. Int Soc Sci J50:17–28
Udovyk O, Gilek M (2013) Coping with uncertainties in science-based advice informing environmental management of the Baltic Sea. Environ Sci Policy 29:12–23
Udovyk O, Gilek M (2014) Participation and post-normal science in practice? Reality check for hazardous chemicals management in the European marine environment. Futures 63:15–25
UN ECE (1998) Convention on access to information, public participation in decision-making and access to justice in environmental matters. Aarhus, 25 June 1998
van Asselt M, Renn O (2011) Risk governance. J Risk Res 14(4):431–449
Wagner RE (2005) Self governance, polycentrism and federalism: recurring themes in Vincent Ostrom's scholarly oeuvre. J Econ Behav Organ 57(2):173–188
Walther Y, Möllmann C (2014) Bringing integrated ecosystem assessments to real life: a scientific framework for ICES. ICES J Mar Sci 71:1183–1186
Weingart P (1999) Scientific expertise and political accountability: paradoxes of science in politics. Sci Public Policy 26:151–161
Wilson DC (2009) The paradoxes of transparency: science and the ecosystem approach to fisheries management in Europe. Amsterdam University Press, Amsterdam

Part I
Interdisciplinary Case Studies of Environmental Governance

Chapter 2
Eutrophication and the Ecosystem Approach to Management: A Case Study of Baltic Sea Environmental Governance

Mikael Karlsson, Michael Gilek, and Cecilia Lundberg

Abstract This study investigates if and how present institutional structures and interactions between scientific assessment and environmental management are sufficient for implementing the ecosystem approach to management (EAM) in the case of Baltic Sea eutrophication. Concerning governance structures, a number of institutions and policies focus on issues relating to eutrophication. In many cases, the policies are mutually supportive rather than contradictory, as seen, for example, in the case of the mutually supportive BSAP and MSFD. The opposite is true, however, when it comes to the linkages with some other policy areas, in particular regarding agricultural policy, where the EU CAP subsidises intensive agriculture with at best minor consideration of environmental objectives, thereby undermining EAM. Enhanced policy coherence and stricter policies on concrete measures to combat eutrophication seem well needed in order to reach stated environmental objectives. When it comes to assessment-management interactions, the science-policy interface has worked well in periods, but the more specific that policies have become, for example, in the BSAP case, the more question marks have been raised about science by affected stakeholders. At present, outright controversies exist, and EAM is far from realised in eutrophication policy in the Baltic Sea region. Besides coping with remaining uncertainties by improving the knowledge on problems and solutions – not least in terms of the socio-economic impacts of eutrophication – it may therefore be valuable to develop venues for improved stakeholder participation.

Keywords Institutions • Science-policy studies • Marine strategy framework directive • Baltic Sea action plan • Common agricultural policy

---

M. Karlsson (✉) • M. Gilek
School of Natural Sciences, Technology and Environmental Studies, Södertörn University, 14189 Huddinge, Sweden
e-mail: mikael.karlsson@2050.se; michael.gilek@sh.se

C. Lundberg
Department of Biosciences, Environmental and Marine Biology, Åbo Akademi University, 20500 Åbo/Turku, Finland
e-mail: cecilia.lundberg@abo.fi

M. Gilek et al. (eds.), *Environmental Governance of the Baltic Sea*, MARE Publication Series 10, DOI 10.1007/978-3-319-27006-7_2

2.1 Introduction

Anthropogenic nutrient over-enrichment is one of the oldest environmental problems. It has escalated during the last century with the exponential increase of human population and consumption, and today eutrophication is a global problem (Díaz and Rosenberg 2011; Rockström et al. 2009; Wassmann and Olli 2006). Eutrophication in marine systems is well described, both for the global and the regional level (Boesch 2002; Jørgensen and Richardson 1996; Wassmann and Olli 2006).

Eutrophication can be defined as an increased input of nutrients or organic matter into an ecosystem, resulting in an increase in primary production (Nixon 1995, 2009). Key indicators of aquatic eutrophication include increases in the total amount of phosphorus and nitrogen, chlorophyll and decreased water transparency. The primary effects are increased production of filamentous algae, changed species composition of microalgae and an increased probability for harmful, and potentially toxic, algal blooms. A complex array of secondary effects may also occur, for example, oxygen deficiency and poorer living conditions for perennial underwater vegetation, immobile zoobenthos living in bottom sediments and certain fish species (Fig. 2.1). These may in turn amplify ecological and associated socio-economic impacts, potentially impeding recovery processes (Lundberg 2005). Extended hypoxic (low oxygen saturation) or anoxic (complete oxygen deficiency) bottom areas (so-called dead zones) are key resultant stressors in marine ecosystems, and the Baltic Sea is the largest stressed ecosystem in the world in this respect (Carstensen et al. 2014; Díaz and Rosenberg 2008).

The first signs of eutrophication on a larger scale in the Baltic Sea became apparent in the 1960s, when oxygen deficiency in the central area was linked to human activities (Elmgren 2001; Fonselius 1969; Jansson 1997; Lundberg 2014). The main anthropogenic sources of eutrophication are agriculture (including crop cultivation and animal husbandry), industries, municipal sewage water and atmospheric deposition (Elmgren and Larsson 2001; HELCOM 2009a, 2013; Wassmann and Olli 2006). However, there was a time lag of two decades after the initial findings in the 1960s, before the issue generated broader public awareness. Today, though, the network of organisations working for the protection and restoration of the marine environment is well developed in the countries surrounding the Baltic Sea (Kern and Löffelsend 2004; Lundberg 2013).

2.1.1 *Governance of Baltic Sea Eutrophication and the Aims of the Study*

Eutrophication, along with overfishing and the presence of hazardous chemicals, constitutes the most serious environmental problems and risks in the Baltic Sea, posing severe threats to biodiversity as well as to other ecosystem services such as

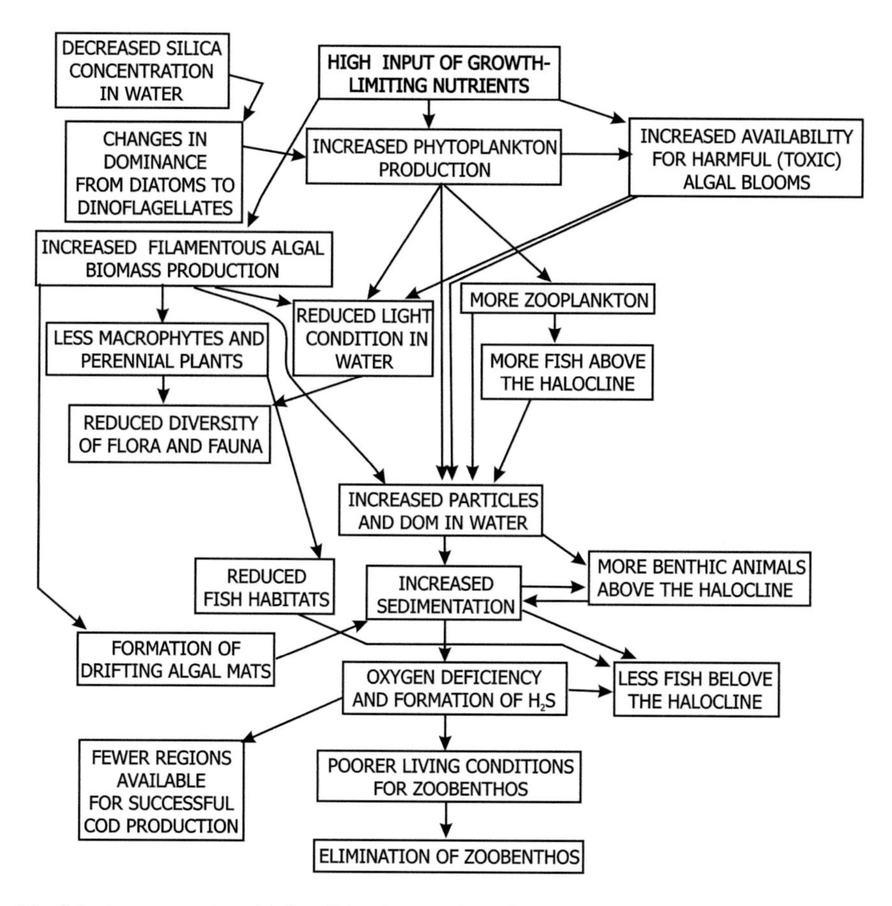

Fig. 2.1 A conceptual model describing the complex primary and secondary ecosystem effects of eutrophication in a marine area

commercial fish and nutrient cycling (HELCOM 2009a, 2010; Korpinen et al. 2012; UNEP 2005). The consequences of eutrophication in the Baltic Sea may ultimately be an ecological regime shift – i.e. a situation where the ecosystem shifts from one relatively stable state to another with unknown thresholds between the states – with associated risks of detrimental effects on ecosystem services and an unknown degree of resilience (Österblom et al. 2010; Österblom et al. 2007).

Traditionally, the governance approach toenvironmental problems and risks has presupposed that various phenomena in natural systems can rationally be dealt with by straightforward science-based management (Linke et al. 2016). The basis for this view has been a reductionist scientific approach, within an international framework of environmental governance, fragmented into sectors and countries and divided into national and international levels (Mee 2005). For example, fisheries and agricultural management have been based on sector-specific laws, policies and institutions (e.g. Sielke and Dreyer in 2015; Mee 2005; Mee et al. 2008), and their relationship to other sectors has seldom been considered.

However, eutrophication is a complex phenomenon. Apart from the intricate array of primary and secondary ecological impacts shown in Fig. 2.1, nutrient sources are diverse and stem from numerous natural and anthropogenic sources in several sectors, both in the drainage basin in question and on a wider international scale through atmospheric and riverine transport (Lundberg 2005, 2014). The relationships between these socio-economic pressures and the marine ecological state are usually non-linear (Mee 2005) and prone to quick and fundamental shifts when thresholds are passed (e.g. Österblom et al. 2010). Furthermore, it may take a long time before any reductions in nutrient input from land may allow for an improved situation (Elofsson 2010). The degradation of organic matter can, for example, be inhibited by negative effects of existing hypoxia on infauna (sediment living animals) (Conley et al. 2007) or by so-called internal loading if buried phosphorus leaks from anoxic sediments (Vahtera et al. 2007; Zillén et al. 2008). In addition, other major disturbances, such as overfishing, and introduction of invasive species may also influence the recovery of an ecosystem from eutrophication. Variability in climate-related factors, such as storms and water temperature and stratification, may additionally cause unexpected responses. The interwoven links between marine eutrophication and all these natural as well as human-induced, biotic as well as abiotic, processes, systems and feedback mechanisms (Caddy 1993; Cloern 2001; McQuatters-Gollop et al. 2009) make science-based reductionist governance models highly insufficient. As Elliot (2002) has pointed out, the management of marine ecosystems needs to consider this full complexity.

In response to these shortcomings, the ecosystem approach to management (EAM) has emerged during the last decades as a central component of environmental governance (Atkins et al. 2011; Curtin and Prellezo 2010; Trush and Dayton 2010). In the words of Browman and Stergiou (2004), EAM is based on the insight that the whole complex ecosystem (including its capacity to deliver important ecosystem services) is greater than the sum of its parts. Problems and risks need to be managed in a holistic manner, not independent of each other (Hammer 2015), and both ecological and social dimensions need to be considered. Moreover, EAM takes into account existing knowledge as well as uncertainties and other forms of complexity. Besides for being different because it is based on a multiple factor approach, EAM also differs from traditional management in its application to a specific geographic scale (Curtin and Prellezo 2010).

EAM is defined in the Convention for Biological Diversity (CBD 1998, 2004) as well as in conventions on regional seas, for example, those concerning the governance of the North Sea and the Baltic Sea (HELCOM and OSPAR 2003). It is also included in several policies and laws, for example, the EU Water Framework Directive (WFD) (EC 2000), the Marine Strategy Framework Directive (MSFD) (EC 2008) and the Helsinki Commission's (HELCOM) Baltic Sea Action Plan (BSAP) (HELCOM 2007a).

Implementation of EAM in the case of eutrophication requires comprehensive knowledge on how both ecosystems and sociopolitical systems function. It remains to be seen, however, whether or not the institutions, policies, action plans and measures in place today in the Baltic Sea region are sufficiently developed to do so in order to promote and reach agreed environmental targets. With that question in mind, this study describes and analyses (1) the formal institutional structures and (2) the interactions between scientific assessment and environmental management in the case of eutrophication in the Baltic Sea region. The aim is to investigate how and to what extent public governance structures and policies take on and implement EAM and what that might mean in relation to environmental policy objectives in place. Furthermore, the study focuses on the question of assessment-management interactions with respect to knowledge integration and the way uncertainty and disagreement is dealt with or not. For practical reasons all atmospheric sources of nitrogen deposition, such as road-based transportation, are not included. Primary focus is placed on the public governance system as it plays the most important role in Baltic Sea environmental governance (Kern 2011).

2.2 Material and Methods

The study is based on both an analysis of documents and a series of interviews. The empirical material presented and analysed consists of various scientific, legal and policy documents as well as of results from qualitative semi-structured interviews with 17 key stakeholders, of which four were active on an international basis and the rest in three countries. Several of those interviewed came from Finland and Sweden, thereby allowing us to assess the situation in two countries well known for comparatively ambitious environmental policies. The interviewees represented four groups: public decision-makers and authorities (six persons), scientists (five persons), NGOs (four persons) and national interest organisations (two persons) (Table 2.1). Public decision-making institutions and authorities were included by persons from the secretariat of the international HELCOM body, ministries of environment and rural affairs in Finland and Sweden and the Swedish Environmental Protection Agency (SEPA). SEPA is partly an independent authority, but it is acting on the basis of Swedish and EU law.

The Baltic Sea environmental scientists came from three universities in Sweden and two research institutes in Denmark and Finland. The non-governmental organisations were both international environmental NGOs and voluntary and politically independent organisations, like the Union of the Baltic Cities (UBC), a network of over 100 cities in the Baltic Sea region. The WWF Sweden, the Coalition Clean Baltic (CCB), the John Nurminen Foundation and the Baltic Sea Action Group (BSAG) were international or national environmental NGOs. The biggest national interest organisations of agriculture in Finland and Sweden, MTK (the Central Union of Agricultural Producers and Forest Owners) and LRF (the Federation of Swedish Farmers), respectively, represented the last stakeholder group.

Table 2.1 The groups of stakeholders interviewed and the organisations and institutions they came from

Group of stakeholder	Organisation	Country
Authorities	HELCOM	International
	Ministry of Environment	Finland
	Ministry of Environment	Sweden
	Ministry of Rural Affairs	Sweden
	Swedish Environmental Protection Agency, SEPA	Sweden
Scientists	Lund University	Sweden
	Stockholm University	Sweden
	Uppsala University	Sweden
	National Environment Research Institute, Aarhus University	Denmark
	Finnish Environment Institute	Finland
NGOs	WWF	International
	Coalition Clean Baltic, CCB	International
	John Nurminen Foundation	Finland
	Baltic Sea Action Group, BSAG	Finland
	Union of the Baltic Cities, UBC	International
National interest organisations	Central Union of Agricultural Producers and Forest Owners, MTK	Finland
	Federation of Swedish Farmers, LRF	Sweden

The empirical work presented in this chapter was performed in 2009–2014 as part of the research projects RISKGOV[1] and COOP.[2] The interviews followed a common guideline from the RISKGOV project (see Gilek et al. 2016) but were especially adapted to suit the key aspects and questions in relation to eutrophication in the Baltic Sea region. The interviews, which lasted approximately between 1 and 2 h, were taped and transcribed, while the interviewed persons were assured anonymity. The analysis of the material followed the empirical methods of qualitative content analysis (Kvale 1996; Silverman 1993).

2.3 Governance Structures Related to Eutrophication

The structure of governance bodies, policies and regulatory frameworks plays an important role in developing and implementing EAM. For example, the level of coordination between institutions and legislation, both vertically and horizontally, influences the effectiveness of risk mitigating measures. On the one hand there might be tensions between top-down management and local influence, and on the

[1] Environmental Risk Governance of the Baltic Sea Studies (2009–2015) www.sh.se/riskgov

[2] Cooperating for Sustainable Regional Marine Governance (2012–2015).

other hand, there might be goal conflicts or synergies at similar levels in the environmental governance landscape. In the following, we will first describe and then analyse the situation in the case of eutrophication in the Baltic Sea.

2.3.1 Key International and Regional Governance Structures

The European Union's importance in the Baltic Sea region has increased significantly over time, not least after the 2004 EU enlargement (Kern and Löffelsend 2008; Kern et al. 2008; Tynkkynen et al. 2014). The EU has a unique capacity to legislate and set supranational demands in the field of environmental protection that are often binding for EU member states. The EU legislation of most importance for mitigating eutrophication in the Baltic Sea consists of WFD and MSFD, as well as the Urban Waste Water Treatment Directive (UWWTD) (EEC 1991a) and the Nitrates Directive (ND) (EEC 1991b).

The 1991 UWWTD focuses on the collection, treatment and discharge of urban wastewater and on the treatment and discharge of wastewater from certain industrial sectors.[3] The ND from the same year instead aims to protect natural ground and surface water quality from nitrate pollution caused by agriculture and is thus one of the key instruments for mitigating eutrophication.

On a broader scale, WFD renewed the EU's water policy in the year 2000, and it covers the protection of inland surface waters, transitional waters, coastal waters and groundwater. The ultimate aim is to prevent further deterioration and to achieve a "good status", which includes eutrophication parameters of all European waters by 2015 by the help of River Basin Management Plans. The watercourses in the EU are divided into separate water units according to the natural geographical and hydrological conditions. MSFD covers the entire marine area, outside the coastal reach of the WFD. MSFD was adopted in 2008 for a more effective protection of the marine environment and as a central component of the then emerging integrated maritime policy of the EU. It thus complements WFD for offshore waters and sets out similar goals ("good environmental status"), albeit with a deadline 5 years later (2020). The implementation rests mainly with member states, which are supposed to collaborate in marine regions, one of them being the Baltic Sea. MSFD is explicitly stated to be based on the ecosystem approach to management.

Last but definitely not least in this context, the EU's Common Agricultural Policy (CAP) is central for the nutrient impact from crop to livestock production (see, e.g. Chen et al. 2014; Schumacher 2012). After a so-called health check in 2008, CAP became a target for contested reform between 2010 and 2013, with one among several aims being "greening the CAP", the latter based on studies showing obvious

---

[3] Related to wastewater treatment, both HELCOM and EU have taken decisions and measures against phosphates in, e.g. detergents, which have led to environmental improvements but also to criticism for filling an "alibi function" and consequently drawing attention away from other more complex issues (see further Schumacher 2012).

shortcomings in relation to environmental objectives (e.g. Baldock et al. 2002). The European Commission indeed presented proposals in that direction (European Commission 2011a), even though they were seen as relatively minor steps (SBA 2012; SRU 2013; Allen and Hart 2013). However, as the regulatory process went on, even these proposals were significantly watered down (European Commission 2012a; IEEP 2013), to the extent that the EU Environment Commissioner stated at the end of the process that he could "only regret that the numerous exemptions, loopholes and thresholds have made the greening so complicated and at the same time have greatly lowered the level of environmental ambition[4]".

In these regulatory contexts, the legislative process is dominated by co-decision, meaning involvement of three large institutions, the European Commission, with its monopoly to present legislative proposals, the Council and the European Parliament, being the key bodies negotiating and jointly adopting the final laws. Much of the implementation, however, rests with the member states, including the national application of the CAP.

Besides legislation, the EU Strategy for the Baltic Sea Region (European Commission 2009a, 2012a) aims at managing the Baltic Sea as a common resource from several perspectives, including environmental issues. The strategy involves all EU members bordering the Baltic Sea and also has the objective to govern EU external issues, such as the relations with the Russian Federation, which contributes significantly to emissions responsible for eutrophication. The original 2009 strategy underlined the need for concrete action instead of new institutions, and linked to the strategy, the European Commission presented an "indicative action plan", which expressed "reduce nutrient inputs" as one of four priorities in the save the sea section of the plan (European Commission 2009b). After evaluation (European Commission 2011b), the strategy was updated and renewed in 2012 (European Commission 2012b).

Turning to the regional Baltic level, the HELCOM operative body of the Helsinki Convention has aimed since the 1970s to promote cooperation between the Baltic Sea states and to monitor and assess the state of the marine environment. HELCOM consists of the parties to the Helsinki Convention of 1974 and is governed both by declarations adopted at infrequent ministerial meetings and by an operative body with an office in Helsinki, under which several staff and expert groups work on, e.g. monitoring and assessment, as well as with drafting proposals on measures and implementation.[5]

Concerning eutrophication, the 1974 Convention contained (Annex 3) specified goals, criteria and measures for preventing land-based pollution, such as emissions of nutrients from sewage water. A central target was then set in a 1988 Ministerial Declaration (HELCOM 1988), which called for a 50 % reduction of nutrient discharges to water and air between 1987 and 1995. In the 1992 renewed version of the Convention, pollution from agriculture was included (Annex 3) as were quite detailed provisions on, for instance, animal density, manure storage and fertilisers.

[4] See, e.g. Janez Potocnik at https://twitter.com/janezpotocnik22/status/350610909284143104

[5] See further at www.helcom.fi

To operationalise the Convention, HELCOM traditionally focuses on adopting more detailed recommendations to the parties of the Convention, specifying, e.g. proposed measures for reaching agreed objectives in the Convention and various declarations, for instance,[6] Recommendation n.b. 28E/4 on plant nutrients (HELCOM 2007b). The binding decisions following agreements within HELCOM are therefore foremost supposed to be taken on a national level, and to be carried out through national implementation programmes (NIPs) (Backer et al. 2010; Tynkkynen et al. 2014), or at the EU level. Since the most central HELCOM agreements are adopted at the ministerial level, i.e. with strong governmental support, a high level of implementation is generally expected, to judge from our interviewees, even though the legislative power as such is weaker than at the EU level. On the other hand, HELCOM, in contrast to the EU, involves all countries in the Baltic Sea region, where the Russian Federation is an important stakeholder from an environmental point of view (Tynkkynen et al. 2014).

The work of HELCOM has changed over time, following the fall of the Berlin Wall, the resultant 1992 amendment of the Helsinki Convention and the enlargement of the EU. More recently, HELCOM has opened more broadly for participation of non-governmental organisations, such as the Swedish and Finnish national agricultural organisations LRF and MTK, and the NGOs WWF and CCB, which all have observer status. Interviewees highlighted that this reflected an important attitude change since stakeholders such as farmers and NGOs generally are seen to play important roles when it comes to eutrophication management.

HELCOM's most central and precise tool at present for marine governance is the 2007 BSAP, which aims to achieve good ecological status in the Baltic Sea by 2021, much in line with WFD and MSFD (Backer and Leppänen 2008; Backer et al. 2010). The BSAP aims to take a broad and systematic approach and defines visions, goals, objectives, indicators, environmental targets and concrete management actions (Backer et al. 2010), which are stated explicitly to be based on EAM and aim to guide the implementation of environmental measures (Backer and Leppänen 2008; Backer et al. 2010; European Commission 2009a; HELCOM 2007a, 2010). In BSAP, national reduction targets for nitrogen and phosphorus are specified.

2.3.2 Analysis of Structural Challenges

As pointed out earlier, the number of governmental bodies and regulatory frameworks related to governance of eutrophication in the Baltic Sea region is high. Governance at the international level dominates and steers the direction of action, with EU legislation and HELCOM action plans and recommendations as the most

[6] HELCOM has adopted a number of Recommendations related to eutrophication, for instance, regarding: wastewater (nb. 6/7), phosphorus (13/10) and nutrients in general (7/2) in agriculture, nitrogen in sewage water (16/17) and agriculture (24/3); see http://helcom.fi/helcom-at-work/recommendations

prominent features. Global organisations such as IMO (the International Maritime Organization) and ICES (the International Council for the Exploration of the Sea), however, also play important roles. The prominent role of international bodies means that governance is significantly a top-down approach.

When it comes to EU legislation, ND and UWWTD both aim at controlling and reducing emissions in the aquatic environment from agriculture and municipalities, respectively. However, even if ND clearly has led to decreased nitrogen emissions from EU agriculture over time (Velthof et al. 2014), it has been considered far from sufficient, influencing, e.g. only a few percent of the nitrogen emissions from manure management (HELCOM 2006b). Similarly, UWWTD requirements are too weak in relation to agreed objectives for the Baltic Sea (Schumacher 2011). Evident shortcomings can be seen also in the case of WFD when it comes to obtaining the good status objective by 2015 (European Commission 2012c). Regarding the MSFD, the implementation is still at an initial stage, but, for example, the definitions of good ecological status for the so-called descriptor 5 for eutrophication ("human-induced eutrophication is minimised, especially adverse effects thereof, such as losses in biodiversity, ecosystem degradation, harmful algae bloom and oxygen deficiency in bottom waters") have been shown to mostly not be adequate (European Commission 2014), including for the Baltic marine region (Milieu 2014).

Coordination of the many governmental bodies and regulatory frameworks at EU, Baltic Sea and national levels is a further key structural challenge. A majority of the interviewees saw significant problems with overlapping roles and ineffectiveness among the actors involved in eutrophication governance in the Baltic Sea region, for example, risks for duplicating similar measures (see also Tynkkynen et al. 2014). Cooperation of multiple actors is required for the implementation of policies to be successful at the regional level (Joas et al. 2007).

The EU Strategy for the Baltic Sea Region (EUSBSR) (European Commission 2009a) is intended to stimulate much needed coordination and cooperation, by integrating different policies and directives in different areas. An important mechanism of the strategy is the promotion of so-called flagship projects with potentially high macro-regional impact in terms of contributing to the fulfilment of agreed objectives. For example, linked to the priority areas of nutrients (PA Nutri), several flagship projects have been initiated, such as Baltic Deal, to promote practices to reduce nutrient losses from agriculture and a flagship project aiming to support the implementation of HELCOM Recommendation 28E/7 on phasing out phosphates in detergents.[7] However, also central to addressing eutrophication would be if the strategy would help to coordinate MSFD and other environmentally focused tools with policies that promote development of the agricultural sector, in particular the referred CAP (European Commission 2011a), but that is so far not done adequately. While the strategy's theoretical structure might seem carefully planned, some interviewees stated that the programme might be too general and vague and too top-down regulated inside the EU. Regarding agriculture, the Commission's action plan to the strategy identifies a need to mitigate nutrient losses, but the strategy and plan

[7] See more about PA Nutri on http://groupspaces.com/eusbsr-nutrient-inputs/

present no proposal on how to restructure CAP – for instance, the present focus on subsidising quite intensive conventional agriculture – in order to promote the marine environmental objectives at hand, not even after the recent CAP reform. There is an obvious lack of integration of the various policies (Baldock et al. 2002; Schumacher 2011; Tynkkynen et al. 2014). It is therefore not surprising that also the interviewees often considered CAP not to be in synergy with MSFD and neither with WFD. Furthermore, communication with Russia is also problematic. In the interviews, opinions were raised that the Russian stakeholders wanted to focus more on cooperation within HELCOM, instead of on EU-related instruments.

Turning to the national level, much of the EU legislation, such as MSFD as well as BSAP allow, presume and sometimes also demand that countries adapt implementation to national circumstances, giving nation states a central role. However, the 2011 National Implementation Plans (NIPs) linked to BSAP diverged quite substantially in terms of content, structure and detail, as well as with respect to implementation of EAM-related measures (Gilek et al. 2013). Only about half of NIPs (e.g. Russia, Estonia and Sweden) gave information on costs and financing (Gilek et al. 2013), and not all countries described planned projects and measures. Regarding EAM implementation, several countries including Denmark, Finland and Russia did not explicitly mention EAM in their NIPs, whereas others referred to EAM principles in a non-specific way. It was also striking that multi-sector cooperation, stakeholder participation and communication, as well as adaptive governance generally were given limited attention. These NIPs revealed that key aspects of EAM have hardly been implemented in national governance frameworks to tackle eutrophication. The dominance of end-of-pipe methods for reducing nutrients, such as wastewater treatment, is increasingly expensive and hard to expand further and therefore insufficient for achieving the significant reductions needed for fulfilling BSAP and MSFD goals.

2.4 Assessment and Management of Eutrophication

EAM evidently relates to both science and policy. For example, the organisation of risk assessment activities and integration of knowledge across different scientific disciplines, as well as addressing ways to cope with uncertainty and disagreement, are central for EAM implementation. On the one hand, formal assessments established by experts based on solid consensual knowledge play an indispensable role for being able to characterise problems and develop science-based advice on particular measures. On the other hand, scientific uncertainty and sociopolitical ambiguity challenge the conventional view on management and open for controversies that might necessitate not only precautionary approaches but also institutionalising deliberative forums. In the following, we will first describe and then analyse the situation in the case of eutrophication in the Baltic Sea.

2.4.1 Key Assessment-Management Interactions

Scientific knowledge and science-based advice have at least since the 1970s played a key role in eutrophication management of the Baltic Sea, for example, influencing decisions on which nutrients to prioritise in wastewater treatment (e.g. Elmgren 2001). Over the years, these interactions between science and policy have developed and changed under the influence of, for example, changes in the knowledge base vis-à-vis remaining uncertainties, stakeholder and public perceptions of eutrophication and associated societal consequences and trade-offs. Interactions have also changed due to transformations of national and international environmental governance arrangements (e.g. Linke et al. 2014). Originally, scientific assessments and science-based advice mainly influenced management measures at the national level. Examples of these are Swedish measures in the 1970–1980s to introduce and expand nitrogen treatment in coastal sewage treatment plants (Elmgren 2001). Successively, collaboration in science and management at the regional Baltic Sea and international (e.g. European) levels increased in importance in response to the large spatial scale of eutrophication impacts and potential solutions in the Baltic Sea. Today the role of science remains strong in the development of policy and management measures, and although the national level is still important for generating environmental assessments linked to various objectives, the regional (i.e. HELCOM) and European (i.e. EU institutions) levels have become the primary domains for science-based policy advice.

When it comes to the regional Baltic Sea level, HELCOM and not the EU is the overall coordinator of the commonly nationally based assessments of eutrophication impacts. In line with ambitions to coordinate national assessment activities and implement EAM, HELCOM has developed a new holistic environmental assessment strategy (HOLAS). This strategy involves recurring integrated thematic assessments of, for example, eutrophication and hazardous chemicals, as well as holistic assessments that aim at assessing ecosystem quality and integrating various societal pressures. An initial holistic assessment was published in 2010 (HELCOM 2010), based on the results from thematic assessments of various environmental issues and objectives linked to BSAP and EU Directives, such as WFD and MSFD (Gilek et al. 2015). To facilitate these thematic assessments and to improve methodological harmonisation and data integration possibilities in the region, HELCOM has also developed various tools for assessing, for example, eutrophication (HELCOM Eutrophication Assessment Tool, HEAT) (HELCOM 2006a, 2009a) and biodiversity (HELCOM Biodiversity Assessment Tool, BEAT) (HELCOM 2009b).

Regarding HELCOM's various proposals on management measures, the 2007 BSAP (with updates in 2013), with its acknowledgement of EAM and agreed national reduction targets for nutrients, is of central importance. According to several interviewees, BSAP evolved as a reaction to the perceived failure of the abovementioned HELCOM target of 50 % nutrient reductions, which was not based on any scientific studies on what is needed to reach a desired state of the marine environment. In contrast, BSAP was in line with EAM and based on scientific assessments of ecological indicators that relate to specific policy objectives associated with "good ecological status" (Backer 2008; HELCOM 2007a, 2010). The actual forebearer to the agreed specific reduction targets and their division among the countries was the decision support system Baltic Nest[8] created within a Swedish research programme (MARE, 1999–2006) and run by the Baltic Nest Institute (BNI 2014). The Baltic Nest system integrates environmental data with economic parameters to build scenarios and generate advice to decision-makers in HELCOM. The reason for including the economic dimension is the understanding among HELCOM parties that a harmonised and collective approach to management is the best way to reach positive environmental outcomes. One motive for this is the desire to achieve as cost-effective nutrient reduction as possible (see, e.g. Elofsson 2002).

2.4.2 Analysis of Assessment-Management Challenges

The new holistic HELCOM assessment strategy (HOLAS) has in many respects implied an improved integration of (mainly natural) science knowledge. For example, the strategy involves an improved spatial *integration* of knowledge of the entire Baltic Sea area, and a wider span of measurements and data are now integrated (e.g. nutrient concentrations, biota, water, oxygen levels). However, although the holistic HELCOM strategy does include specific (albeit rather limited) sections discussing social and economic aspects (HELCOM 2010), the integration of social and economic aspects is still rather limited in HELCOM activities. This is discernible at both the assessment and management levels. However, there is a high awareness of this lack of integrated interdisciplinary assessments among decision-makers and scientists alike (as observed in our interviews), and recent efforts by the international research network BalticSTERN to develop, for example, cost-benefit analyses of eutrophication management in the Baltic Sea have started to improve the situation (BalticSTERN 2013). According to these estimates, the societal benefits for reaching BSAP targets would exceed costs (amounting to 2300–2800 million euro, depending on how cost-effective measures are) by 1000–1500 million euros annually. More recent studies have estimated that the costs could be higher (Wulff et al. 2014). Investigations like these are nevertheless important in order to provide ground for better integration of social and natural parameters.

In spite of this, our interviewees underlined several remaining challenges for achieving adequately integrated assessments of eutrophication. These challenges relate to, for example, a substantial shortage of data from some geographic areas and problems in reaching agreement among countries and stakeholders on target levels and thresholds to base thematic assessments on (e.g. Haahti et al. 2010; Lundberg 2013). Consequently, we conclude that, despite recent efforts by HELCOM and BalticSTERN to develop integrated assessments of eutrophication to support

[8] See Baltic Nest at http://www.balticnest.org

EAM, there is still quite a way to go before socio-economic consequences and concerns are sufficiently addressed in assessments and science-based advice.

In terms of *uncertainty challenges*, both BSAP and MSFD do, in line with EAM, refer to ecosystem complexity and the importance of applying a precautionary approach in marine environmental governance (cf. Udovyk and Gilek 2014). For example, HELCOM and OSPAR (2003) define EAM in the marine environment as:

> [....] the comprehensive integrated management of human activities based on the best available scientific knowledge about the ecosystem and its dynamics, in order to identify and take action on influences which are critical to the health of marine ecosystems, thereby achieving sustainable use of ecosystem goods and services and maintenance of ecosystem integrity. [...] The application of the precautionary principle is equally a central part of the ecosystem approach.

However, studies of guideline, assessment and advice documents linked to MSFD (Udovyk and Gilek 2014) and the HELCOM BSAP (Udovyk and Gilek 2013) reveal a rather limited acknowledgement and management of uncertainty. In fact, irreducible uncertainties associated with ecosystem dynamics and interdependencies are rarely mentioned in assessments and science-based advice (Udovyk and Gilek 2013). Similarly, there are hardly any references to strategies or methods for coping with such uncertainty. Instead, in line with the notion of achieving "best available knowledge", assessment and advice documents mainly acknowledge uncertainty caused by low precision and accuracy in methods and a general lack of scientific data for certain geographical areas and ecological endpoints.

The general strategy applied for managing epistemic uncertainty is to obtain more data through an expanded monitoring network, with larger geographic and temporal coverage (HELCOM 2009a). Such a traditional empirical approach in science has in many ways been successful in and instrumental to reaching a consensual understanding of the sources and impacts of eutrophication and of the importance of various nutrients (e.g. Conley et al. 2009a; Elmgren 2001). However, for generating science-based advice on nutrient reduction requirements to reach environmental objectives, alternative modelling approaches would be needed to better control uncertainty associated with ecosystem dynamics (e.g. Udovyk and Gilek 2013). Interestingly, however, our interviews revealed a not so uncommon "downplaying" of model and scenario uncertainties in science-policy interactions linked to development of BSAP. Presumably, this can to some extent be explained by a common ambition among scientists (Baltic Nest) and decision-makers (HELCOM) involved to facilitate a regional agreement on nutrient reduction targets (cf. Linke et al 2014). A strong acknowledgement of uncertainty could in this respect have been a reason for disagreement rather than agreement.

An overview of scientific studies on Baltic Sea eutrophication exposes several scientific *disagreements* on the sources and impacts of eutrophication that in various ways have had significant repercussions on stakeholder conflicts and management decisions at national and regional levels (e.g. Elmgren 2001). At an early stage of Baltic-wide eutrophication assessment and management, there was in the 1960s a lively debate on whether or not anthropogenic eutrophication of the open Baltic Sea was possible at all. Once compelling evidence for such large-scale human-induced eutrophication was available, there was instead an infected debate whether nitrogen or phosphorus is the main causative nutrient in need of reduction measures (e.g. Elmgren 2001).

Today, these natural science-based disagreements on nutrients have, according to the interviewed scientists, been handled, and quite a consensual scientific view exists, that emissions and levels of both nitrogen and phosphorus need to decrease (Conley et al. 2009a). However, some scientific disagreements still exist, mainly with regard to methodological details, but still also according to our interviewees to which nutrient to preferentially reduce.

As mentioned above, the process to develop and agree on the 2007 BSAP was associated with a rather surprisingly low level of disagreement among stakeholders. This consensual assessment-management process has been attributed to the close cooperation between the Baltic Nest Institute and HELCOM while developing science-based advice (Linke et al. 2014). For example, it has been argued that the "HELCOM-Nest nexus" demonstrates how scientific assessments and science-based advice may underpin the legitimacy of political claims for regional environmental management (Linke et al. 2014). However, interviews with various actors revealed that the interplay between scientific data used in the Baltic Nest system and HELCOM's management responses is interpreted differently by stakeholders in the Baltic Sea region. Some see the Baltic Nest system as a concrete and illustrative tool for coming up with effective remedies, while others are more critical and argue that the model has received too much attention at the expense of other models (cf. Linke et al. 2014). Recently, strong criticism from, for example, farmers has also been voiced concerning the political conclusions on the sharing of responsibility for national reduction obligations (BFFE 2013; Linke et al. 2014; LRF 2013). The Baltic Nest model is, however, under continuous development and improvement (BNI 2014; Johansson et al. 2007), and the linkage between risk assessment and risk management remains central to the governance of eutrophication in the Baltic Sea.

What also came up in interviews is that there is an ongoing discussion, linked to implementation of eutrophication policy, in the scientific literature on the potential effectiveness and efficiency of various management measures. For example, there are disagreements concerning which specific reduction measures are most cost-effective when comparing costs for direct emission reduction measures with costs for land-use changes designed to increase nutrient retention (e.g. Elofsson 2010; Gren 2008; Huhtala et al. 2009; Lundberg 2013). Some studies and experiments have also suggested and tested alternative, technical solutions to reduce nutrient concentrations in the Baltic Sea such as artificial oxygenation, changes in saltwater inflow and chemical sequestration of phosphorus buried in the sediment (see, e.g. the review by Conley et al. 2009b). Still, although some "engineering"-type measures, such as phosphorus binding with aluminium, were argued to potentially be effective in specific coastal areas, their potential to address open sea eutrophication was generally seen as marginal by the scientists, NGOs and decision-makers we interviewed. The overall view is that external nutrient reduction – i.e. before nutrients enter the sea – is the only truly effective long-term strategy to combat eutrophication (as, e.g. also argued by Conley et al. 2009a).

2.5 Conclusions and Recommendations

This study has described and analysed if and how present institutional structures and interactions between scientific assessment and environmental management are sufficient for implementing EAM in the case of Baltic Sea eutrophication and what that might mean in relation to the policy objectives in place. We have also studied knowledge integration and the way uncertainty and disagreement is dealt with or not when it comes to assessment-management interactions.

Regarding *governance structures*, we have shown that there is a wide array of eutrophication-related policies and institutions in place at national, Baltic Sea and European levels. Obviously there is a risk that such complex structures may imply contradictory, overlapping or redundant institutional arrangements which might lead to institutional tensions and inefficiencies (cf. Tynkkynen et al. 2014). However, we have in this case study observed mostly synergistic institutional interactions to deal with Baltic Sea eutrophication, where policies focused on mitigating eutrophication seem to mutually enforce each other. A striking example here is MSFD and BSAP where nothing in BSAP prevents implementing MSFD and vice versa. On the contrary, the eutrophication segments of these policies seem to have developed in a rather co-evolutionary manner – both are, for example, explicitly based on EAM and a Baltic Sea-wide coordination of management measures. These synergies support the implementation of MSFD and BSAP and may even strengthen the regulatory weaker HELCOM's position and role in eutrophication governance, especially since Russia is a party to the Helsinki Convention (cf. Söderström et al. 2015).

However, there are also obvious conflicting institutional interactions linked to eutrophication, particularly between sectoral (e.g. agriculture) and environmental policies in the EU system (cf. Tynkkynen et al. 2014; De Santo 2015). Most important for this context, the CAP continues to subsidise intensive agriculture with at best only minor consideration of environmental objectives for the marine environment, allowing for high inputs of nutrients, of which much ends up in the Baltic Sea sooner or later. In spite of EAM being expressed in MSFD, and in spite of the aim of EUSBSR, no effective coordination mechanism in relation to agriculture is in place today.

In order to enhance possibilities to reach the eutrophication objective in place, a number of structural changes seem warranted:

First, multi-sector coordination of policies is needed within European, regional and national institutional structures. This is most obvious within the EU, where goals and means in the EU's CAP should be adequately adapted to EU's environmental objectives (as, e.g. manifested in the MSFD), in line with the environmental integration principle in the treaty (TFEU 2007, Article 11). Without EU regulatory harmony between the policy domains of agriculture and environment, EAM will hardly be applied in a relevant manner in reality. Second, further multilevel coordination of objectives and policies adopted by institutions at European, Baltic Sea and national levels is possible, for example, between the EU and HELCOM. HELCOM, and the EU, should strive to create and strengthen synergies with each other's policies based on EAM. We see, for example, no reason why the objectives, timetables and programmes should differ between, for example, MSFD and BSAP. Finally, effective concrete measures and financing for these are, at the end of the day, always indispensable, irrespective of the institution and regulatory level. Several ideas for potentially positive concrete measures also came up in this study, not least in the interviews, for example, regarding the short-term need for reducing the use and, thereby, the losses of fertilisers and nutrients, respectively, and the long-term need for more fundamental structural changes in agriculture, for instance, by changing the geographical balance between husbandry and crop production. The latter would most likely necessitate a major CAP reform, though, as well as comprehensive national strategies.

When looking at *assessment-management* (*science-policy*) *interactions* in the process leading up to BSAP, what is clear is that this was characterised by a rather straightforward translation of results from the Baltic Nest system to science-based advice and subsequent decisions. This process seems to have been facilitated by a tight coevolutionary interplay focussing on consensual knowledge between scientists linked to the Baltic Nest Institute and those involved in the management regime under HELCOM. During this initial stage of BSAP development, scientific uncertainty linked to eutrophication assessments and advice was not a primary issue of concern, and hardly any major disagreements among either countries or stakeholder groups could be observed. Today, however, during the ongoing national implementation of BSAP, engagement and critique, not least by some farmer's organisations, on eutrophication management strategies and measures, have grown and become far more detailed in terms of, for example, uncertainties and which measures to optimally take, in particular in relation to interpretations of cost-benefit analyses. Such increased stakeholder engagement and disagreement in response to implementation of proposed concrete nutrient reduction measures is probably what is to be expected given that different stakeholders' values and interests are related to different costs and benefits during implementation.

Our analyses of science-policy interactions also reveal that significant challenges still remain in terms of elaborating concrete strategies for implementing EAM, which we argue is needed to reach a good environmental status in terms of eutrophication in the Baltic Sea:

First, integration of various forms of knowledge relating to social, economic and environmental risks, costs and benefits of eutrophication is indispensible for implementing EAM. However, despite a general awareness of this need among decision-makers, scientists and other stakeholders, and despite recent substantial contributions by the BalticSTERN research network, socio-economic knowledge, assessments and advice on eutrophication are still in need of development. One set of issues far from resolved concerns how to optimally allocate responsibilities for reducing nutrient loads in line with BSAP and if optimal means cost-effective (by means of, e.g. cap and trade) or something else (e.g. BalticStern 2013; Ahtiainen et al. 2014; Tynkkynen et al. 2014; Wulff et al. 2014). Second, given the complex ecosystem dynamics associated with eutrophication, coping with fundamental uncertainties is a basic challenge when striving to implement EAM in the case of Baltic Sea eutrophication governance (cf. Österblom 2010). Finally, realising ambitions of developing integrated science-based advice on effective concrete measures will expose different norms, principles, methodologies, assumptions, etc. in different academic traditions and might potentially lead to increased levels of scientific – and in turn, broader – disagreement, at least in the short run.

In relation to these challenges and how they have been handled by the science system, it is interesting to note that, apart from the Baltic Nest Institute, the arenas for consensus building in the case of eutrophication have been rather diffuse, including a variety of HELCOM groups, projects and national review groups. Therefore, we consider (as also argued by some of our interviewed scientists) that improved regional integration and coordination of eutrophication-related science is needed, perhaps as a permanent independent Baltic Sea Science Panel that recurrently could review the state of science on environmental issues in the Baltic Sea and develop guidance on modelling and scenarios as well as on science-based approaches to better cope with knowledge integration and uncertainty. For example, linked to coping with fundamental uncertainties in science-based advice, precautionary strategies could be developed based on a combination of approaches and methodologies already published in the academic literature (Udovyk and Gilek 2013), for example, default factors and alternative principles for decision-making (cf. Karlsson 2005). In addition, some of our interviewees were of the opinion that it would be possible to learn from approaches for uncertainty appraisal developed by the Intergovernmental Panel on Climate Change (IPCC), to better cope with uncertainty in Baltic Sea eutrophication advice. It was also argued that one of the major advantages with IPCC's work with scenarios and related uncertainties is the number of independent institutions involved, which give the scenarios a certain legitimacy and credibility. Similarly, another potential learning point from the climate discourse would be to develop a nutrient cap and trade system for the Baltic Sea, which could be both goal and cost-effective.

Finally, in our summation of Baltic Sea eutrophication governance, we conclude that policy prescriptions in place are ambitious and promising and that the general knowledge base is quite well developed. Based on this, much has happened in order to mitigate eutrophication, but we can clearly see that several fundamental challenges remain in order to implement EAM and to ultimately reach the overall policy objectives in, for example, BSAP and MSFD. In terms of governance structure, there is primarily a need to improve coordination of agricultural and environmental policies and develop science-policy interactions in line with EAM, where it is vital that interdisciplinary integration and strategies for coping with uncertainty are improved.

Acknowledgments We thank the co-authors of the RISKGOV research report on eutrophication, which formed an important point of departure for this study: Britt-Marie Haahti, Eva Hedenström, Sebastian Linke, Gunilla Reisner and Markus Wanamo. All interviewees are thanked for their participation and valuable contributions. The work was funded by the Foundation for Baltic and East European Studies and the European Community's Seventh Framework Programme (2007–

2013) under the grant agreement no. 217246 made with the joint Baltic Sea research and development programme BONUS, as well as by the Swedish Environmental Protection Agency, the Swedish Research Council Formas and the Academy of Finland.

References

Ahtiainen H, Artell J, Elmgren R, Hasselström L, Håkansson C (2014) Baltic Sea nutrient reductions. What should we aim for? J Environ Manage 145:9–23

Allen B, Hart K (2013) Meeting the EU's environmental challenges through the CAP – how do the reforms measure up? Asp Appl Biol 118:9–22

Atkins JP, Burdon D, Elliott M, Gregory AJ (2011) Management of the marine environment: integrating ecosystem services and societal benefits with the DPSIR framework in a systems approach. Mar Pollut Bull 62:215–226

Backer H (2008) Indicators and scientific knowledge in regional Baltic Sea environmental policy. ICES J Mar Sci 65:1398–1401

Backer H, Leppänen JM (2008) The HELCOM system of a vision, strategic goals and ecological objectives: implementing an ecosystem approach to the management of human activities in the Baltic Sea. Aquat Conserv 18:321–334

Backer H, Leppänen JM, Brusendorff AC, Forsius K, Stankiewicz M, Mehtonen J, Pyhälä M, Laamanen M, Paulomäki H, Vlasov N, Haaranen T (2010) HELCOM Baltic Sea Action Plan – a regional programme of measures for the marine environment based on the ecosystem approach. Mar Pollut Bull 60:642–649

Baldock D, Dwyer J, Sumpsi-Vinas JM (2002) Environmental integration and the CAP. A report to the European Commission, DG Agriculture. IEEP, Brussels

BalticSTERN (2013) The Baltic Sea – our common treasure: economics of saving the sea. Report 2013:4. Swedish Agency for Marine and Water Management, Göteborg. Available from: http://stockholmresilience.org

BFFE (2013) Viewpoints from the farmer organisations around the Baltic Sea to the proposal for the ministerial declaration concerning revised HELCOM Baltic Sea Action Plan (BSAP). Available from: http://www.lrf.se

BNI (Baltic Nest Institute) (2014). Available from: http://www.balticnest.org/. Accessed 7 Nov 2014

Boesch DF (2002) Challenges and opportunities for science in reducing nutrient over-enrichment of coastal ecosystems. Estuaries 25(4B):886–900

Browman HI, Stergiou KI (2004) Perspectives on ecosystem-based approaches to the management of marine resources. Mar Ecol Prog Ser 274:269–303

Caddy JF (1993) Towards a comparative evaluation of human impacts on fishery ecosystems of enclosed and semi-enclosed seas. Rev Fish Sci 1:57–95

Carstensen J, Conley DJ, Bonsdorff E, Gustafsson BG, Hietanen S, Janas U, Jilbert T, Maximov A, Norkko A, Norkko J, Reed DC, Slomp CP, Timmermann K, Voss M (2014) Hypoxia in the Baltic Sea: biogeochemical cycles, benthic fauna, and management. AMBIO 43:26–36

CBD (Convention on Biodiversity) (1998) Report of the workshop on the ecosystem approach. Lilongwe, Malawi, 26–28 Jan 1998. UNEP/COP/4/Inf.9
CBD (Convention on Biodiversity) (2004) The ecosystem approach. Secretariat of the Convention on Biodiversity (CBD), CBD, Montreal
Chen Q, Kozar O, Li F, Pekonen A, Saarman P (2014) Eutrophication in the Baltic Sea – characteristics and challenges. Paper presented at HENVI Science Days, University of Helsinki, 13–14 May 2014
Cloern JE (2001) Our evolving conceptual model of the coastal eutrophication problem. Mar Ecol Prog Ser 210:223–253
Conley DJ, Carstensen J, Ærtebjerg G, Christensen PB, Dalsgaard T, Hansen JLS, Josefson AB (2007) Long-term changes and impacts of hypoxia in Danish coastal waters. Ecol Appl 17(5 Supp):165–184
Conley DJ, Paerl HW, Howarth RW, Boesch DF, Seitzinger SP, Havens KE, Lancelot C, Likens GE (2009a) Controlling eutrophication: nitrogen and phosphorus. Science 323(5917):1014–1015
Conley DJ, Bonsdorff E, Carstensen J, Destouni G, Gustafsson BG, Hansson LA, Rabalais NN, Voss M, Zillén L (2009b) Tackling hypoxia in the Baltic Sea: is engineering a solution? Environ Sci Technol 43:3407–3411
Curtin R, Prellezo R (2010) Understanding marine ecosystem based management: a literature review. Mar Policy 34:821–830
De Santo E (2015) The Marine Strategy Framework Directive as a catalyst for maritime spatial planning: internal dimensions and institutional tensions. In: Gilek M, Kern K (eds) Governing Europe's marine environment. Europeanization of regional seas or regionalization of EU policies? Ashgate Publishing, Farnham
Díaz RJ, Rosenberg R (2008) Spreading dead zones and consequences for marine ecosystems. Science 321(5891):926–929
Díaz RJ, Rosenberg R (2011) Introduction to environmental and economic consequences of hypoxia. Int J Water Resour Dev 27:71–82
EC (2000) Water Framework Directive, 2000/60/EC. OJ L 327, pp 1–72
EC (2008) Marine Strategy Framework Directive, 2008/56/EC. OJ L 164, pp 19–40
EEC (1991a) Urban Wastewater Treatment Directive, 91/271/EEC. OJ L 135, pp 40–52
EEC (1991b) Nitrates Directive, 91/676/EEC. OJ L 375, pp 1–8
Elliott M (2002) The role of the DPSIR approach and conceptual models in marine environmental management: an example for offshore wind power. Mar Pollut Bull 44(6):iii–vii
Elmgren R (2001) Understanding human impact on the Baltic ecosystem: changing views in recent decades. AMBIO 30:222–231
Elmgren R, Larsson U (2001) Eutrophication in the Baltic Sea area: integrated coastal management issues. In: von Bodungen B, Turner RK (eds) Science and integrated coastal management. Dahlem University Press, Berlin, pp 15–35
Elofsson K (2002) Economics of marine pollution. Dissertation, Swedish University of Agricultural Sciences
Elofsson K (2010) The costs of meeting the environmental objectives for the Baltic Sea: a review of the literature. AMBIO 39:49–58
European Commission (2009a) European Union Strategy for the Baltic Sea Region. Communication from the Commission to the European Parliament, the Council, the European Economic and Social Committee and the Committee of the Regions. COM(2009) 248 FINAL
European Commission (2009b) Commission staff working document accompanying the Communication from the Commission to the European Parliament, the Council, the European Economic and Social Committee and the Committee of the Regions concerning the European Union Strategy for the Baltic Sea Region. SEC(2009) 712/2
European Commission (2011a) CAP reform – an explanation of the main elements. MEMO/11/695
European Commission (2011b) Report from the Commission to the European Parliament, the Council, the European Economic and Social Committee and the Committee of the Regions on the implementation of the EU Strategy for the Baltic Sea Region (EUSBSR). COM(2011) 382 FINAL

European Commission (2012a) Concept paper – May 2012, Agricultural Council – greening
European Commission (2012b) Communication from the Commission to the European Parliament, the Council, the European Economic and Social Committee and the Committee of the Regions concerning the European Union Strategy for the Baltic Sea Region. COM(2012) 128 FINAL
European Commission (2012c) Report from the Commission to the European Parliament and the Council on the implementation of the Water Framework Directive (2000/60/EC). River basin management plans. COM(2012) 670 final
European Commission (2014) Commission staff working document. Annex accompanying the document Commission Report to the Council and the European Parliament the first phase of implementation of the Marine Strategy Framework Directive (2008/56/EC). The European Commission's assessment and guidance. COM(2014) 97 Final
Fonselius SH (1969) Hydrography of the Baltic deep basins. III. Fishery Board of Sweden. Ser Hydrogr 23:1–97
Gilek M, Hassler B, Engkvist F, Kern K (2013) The HELCOM Baltic Sea Action Plan: challenges of implementing an innovative ecosystem approach. In: Henningsen B, Etzold T, Pohl AL (eds) Political state of the region report 2013 – trends and directions in the Baltic Sea region. Baltic Development Forum, Copenhagen, Available from: http://www.bsr2013.eu
Gilek M, Karlsson M, Udovyk O, Linke S (2015) Science and policy in the governance of Europe's marine environment – the impact of Europeanization, regionalization and the ecosystem approach to management. In: Gilek M, Kern K (eds) Governing Europe's marine environment. Europeanization of regional seas or regionalization of EU policies? Ashgate Publishing, Farnham
Gilek M, Karlsson M, Linke S, Smolarz K (2016) Environmental governance of the Baltic Sea – identifying key challenges, research topics and analytical approaches. In: Gilek M et al (eds) Environmental governance of the Baltic Sea. Springer, Dordrecht
Gren IM (2008) Cost effectiveness and fairness of the HELCOM Baltic Sea action plan against eutrophication. Vatten 64:273–281
Haahti BM, Hedenström E, Linke S, Lundberg C, Reisner G, Wanamo M (2010) Case-study report eutrophication. Deliverable 2, RISKGOV project, http://www.sh.se/riskgov
Hammer M (2015) The ecosystem management approach. Implications for marine governance. In: Gilek M, Kern K (eds) Governing Europe's marine environment. Europeanization of regional seas or regionalization of EU policies? Ashgate Publishing, Farnham
HELCOM (1988) Declaration on the protection of the marine environment of the Baltic Sea, adapted on 15 February 1988 in Helsinki by the Ministers responsible for the environmental protection in the Baltic Sea states
HELCOM and OSPAR (2003) Statement on the ecosystem approach to the management of human activities. First joint ministerial meeting of the Helsinki and OSPAR Commissions, Bremen, 25–26 June 2003
HELCOM (2006a) Development of tools for assessment of eutrophication in the Baltic Sea. Baltic Sea Environmental Proceedings No. 104. Available from: http://helcom.fi
HELCOM (2006b) Eutrophication in the Baltic Sea; Draft HELCOM thematic assessment in 2006. Baltic Marine Environment Protection Commission, 27th Meeting, Helsinki, Finland, 8–9 March 2006
HELCOM (2007a) HELCOM Baltic Sea Action Plan. Available from: http://helcom.fi
HELCOM (2007b) HELCOM Recommendation 28E/4. Amendments to annex IIII "Criteria and measures concerning the prevention of pollution from land-based sources" of the 1992 Helsinki Convention
HELCOM (2009a) Eutrophication in the Baltic Sea – an integrated thematic assessment of the effects of nutrient enrichment in the Baltic Sea region. Baltic Sea Environmental Proceedings No. 115B
HELCOM (2009b) Biodiversity in the Baltic Sea – an integrated thematic assessment on biodiversity and nature conservation in the Baltic Sea. Baltic Sea Environmental Proceedings No. 116B
HELCOM (2010) Ecosystem health of the Baltic Sea 2003–2007: HELCOM initial holistic assessment. Baltic Sea Environmental Proceedings No. 122

HELCOM (2013) Approaches and methods for eutrophication target setting in the Baltic Sea region. Baltic Sea Environmental Proceedings No. 133

Huhtala A, Ahtiainen H, Ekholm P, Fleming-Lehtinen V, Heikkilä J, Heiskanen, AS, Helin J, Helle I, Hyytiäinen K, Hällfors H, Iho A, Koikkalainen K, Kuikka S, Lehtiniemi M, Mannio J, Mehtonen J, Miettinen A, Mäntyniemi S, Peltonen H, Pouta E, Pylkkö M, Salmiovirta M, Verta M, Vesterinen J, Viitasalo M, Viitasalo-Frösen S, Väisänen S (2009) The economics of the state of the Baltic Sea. Pre-study assessing the feasibility of a cost-benefit analysis of protecting the Baltic Sea ecosystem. Report 2-2009. In: The Advisory Board for Sectoral Research 2:2009. Available from: http://www.minedu.fi

IEEP (Institute for European Environmental Policy) (2013) Environment undermined in CAP deal, Brussels, 26.06.2013

Jansson BO (1997) The Baltic Sea: current and future status and impact of agriculture. AMBIO 26:424–431

Joas M, Kern K, Sandberg S (2007) Actors and arenas in hybrid networks: implications for environmental policymaking in the Baltic Sea region. AMBIO 36:237–242

Johansson S, Bonsdorff E, Wulff F (2007) The MARE research program 1999–2006 – reflections on program management. AMBIO 36:119–122

Jørgensen BB, Richardson K (eds) (1996) Eutrophication in coastal marine ecosystems, vol 52, Coastal and Estuarine Studies. American Geophysical Union, Washington, DC

Karlsson M (2005) Managing complex environmental problems for sustainable development. Academic thesis, Karlstad University Press, Karlstad

Kern K (2011) Governance for sustainable development in the Baltic Sea region. J Balt Stud 42:67–81

Kern K, Löffelsend T (2004) Governance beyond the nation state in the Baltic Sea region. Local Environ 9:451–467

Kern K, Löffelsend T (2008) Governance beyond the nation state: transnationalization and Europeanization of the Baltic Sea region. In: Joas M, Jahn D, Kern K (eds) Environmental policies in the Baltic Sea region. Governing a common sea. Earthscan, London

Kern K, Joas M, Jahn D (2008) Governing a common sea: comparative patterns for sustainable development. In: Joas M, Jahn D, Kern K (eds) Governing a common sea. Environmental policies in the Baltic Sea region. Earthscan, London

Korpinen S, Meski L, Andersen JH, Laamanen M (2012) Human pressures and their potential impact on the Baltic Sea ecosystem. Ecol Indic 15:105–114

Kvale S (1996) Interviews: an introduction to qualitative research interviewing. Sage Publications, London

Linke S, Gilek M, Karlsson M, Udovyk O (2014) Unravelling science-policy interactions in environmental risk governance of the Baltic Sea: comparing fisheries and eutrophication. J Risk Res 17:505–523

Linke S, Gilek M, Karlsson M (2016) Science-policy interfaces in Baltic Sea environmental governance: towards regional cooperation and management of uncertainty? In: Gilek M et al (eds) Environmental governance of the Baltic Sea. Springer, Dordrecht

LRF (2013) Hur återställer vi Östersjön? Effektivare strategier mot ett mindre övergött hav (in Swedish). In: Lantbrukarnas Riksförbund, Stockholm. Available from: http://www.lrf.se

Lundberg C (2005) Conceptualizing the Baltic Sea ecosystem. An interdisciplinary tool for environmental decision-making. AMBIO 34:433–439

Lundberg C (2013) Eutrophication, risk management and sustainability. The perceptions of different stakeholders in the northern Baltic Sea. Mar Pollut Bull 66:143–150

Lundberg C (2014) Water quality of the Baltic Sea. In: Ahuja S (ed) Comprehensive water quality and purification. Elsevier, Waltham, pp 251–269

McQuatters-Gollop A, Gilbert AJ, Mee LD, Vermaat JE, Artioli Y, Humborg C, Wulff F (2009) How well do ecosystem indicators communicate the effects of anthropogenic eutrophication? Estuar Coast Shelf Sci 82:583–596

Mee L (2005) Assessment and monitoring requirements for the adaptive management of Europe's regional seas. In: Vermaat J, Bouwer L, Turner RK, Salomons W (eds) Managing European coasts: past, present and future. Springer, Berlin

Mee LD, Jefferson RL, Laffoley D, Elliott M (2008) How good is good? Human values and Europe's proposed marine strategy directive. Mar Pollut Bull 56(2):187–204

Milieu (2014) Article 12 technical assessment of the MSFD 2012 obligations. Baltic Sea. Final version of report to the European Commission, 7 February 2014

Nixon SW (1995) Coastal marine eutrophication: a definition, social causes, and future concerns. Ophelia 41:199–219

Nixon SW (2009) Eutrophication and the macroscope. Hydrobiologia 629:5–19

Österblom H, Hansson S, Larsson U, Hjerne O, Wulff F, Elmgren R, Folke C (2007) Human-induced trophic cascades and ecological regime shifts in the Baltic Sea. Ecosystems 10:877–889

Österblom H, Gårdmark A, Bergström L, Müller-Karulis B, Folke C, Lindegren M, Casin M, Olsson P, Diekmann R, Blenckner T, Humborg C, Möllmann C (2010) Making the ecosystem approach operational – can regime shifts in ecological- and governance systems facilitate the transition? Mar Policy 34:1290–1299

Rockström J, Steffen W, Noone K, Persson Å, Chapin FS III, Lambin EF, Lenton TM, Scheffer M, Folke C, Schellnhuber HJ, Nykvist B, de Wit CA, Hughes T, van der Leeuw S, Rodhe H, Sörlin S, Snyder PK, Costanza R, Svedin U, Falkenmark M, Karlberg L, Corell RW, Fabry VJ, Hansen J, Walker B, Liverman D, Richardson K, Crutzen P, Foley JA (2009) A safe operating space for humanity. Nature 461:472–475

SBA (Swedish Board of Agriculture) (2012) A greener CAP? An analysis of the Commission's greening proposal for the Common Agricultural Policy. Report 2012:13, Jönköping

Schumacher T (2011) The capacity of the EU to address marine eutrophication. In: Pihlajamäki M, Tynkkynen N (eds) Governing the blue-green Baltic Sea. Societal challenges of marine eutrophication prevention, vol 31, FIIA Report. Finnish Institute of International Affairs (FIIA), Helsinki

Schumacher T (2012) Evaluation of the European Union's policies and legislation related to Baltic Sea eutrophication. PROBALT project. Research Group on International Political Sociology, Institute of Social Sciences, Kiel University, Kiel

Sellke P, Dreyer M, Linke S (2015) Fisheries: a case study of Baltic Sea environmental governance. In: Gilek M et al (eds) Environmental governance of the Baltic Sea. Springer, Dordrecht

Silverman D (1993) Interpreting qualitative data. Methods for analysing talk, text and interaction. Sage Publications, London

Söderström S, Kern K, Hassler B (2015) Marine governance in the Baltic Sea: current trends of Europeanization and regionalization. In: Gilek M, Kern K (eds) Governing Europe's marine environment. Europeanization of regional seas or regionalization of EU policies? Ashgate Publishing, Farnham

SRU (German Advisory Council on the Environment) (2013) Reform of the Common Agricultural Policy: opportunities for reorientation. Comment on environmental policy no 11. Berlin, January 2013

TFEU (2007) Consolidated versions of the treaty on European Union and the treaty on the functioning of the European Union. Off J Eur Union C 326:1–390

Trush SF, Dayton PK (2010) What can ecology contribute to ecosystem-based management? Ann Rev Mar Sci 2:419–441

Tynkkynen N, Schönach P, Pihlajamäki M, Nechiporuk D (2014) The Governance of the mitigation of the Baltic Sea eutrophication: exploring the challenges of the formal governing system. AMBIO 43:105–114

Udovyk O, Gilek M (2013) Coping with uncertainties in science-based advice informing environmental management of the Baltic Sea. Environ Sci Policy 29:12–23

Udovyk O, Gilek M (2014) Participation and post-normal science in practice? Reality check for hazardous chemicals management in the European marine environment. Futures 63:15–25

UNEP (2005) Lääne A, Kraav E, Titova G (eds) Baltic Sea, GIWA, Regional Assessment 17. University of Kalmar, Kalmar

Vahtera E, Conley DJ, Gustafsson BG, Kuosa H, Pitkänen H, Savchuk OP, Tamminen T, Viitasalo M, Voss M, Wasmund N, Wulff F (2007) Internal ecosystem feedbacks enhance nitrogen-fixing cyanobacteria blooms and complicate management in the Baltic Sea. AMBIO 36:186–194

Velthof GL, Lesschen JP, Webb J, Pietrzak S, Miatkowski Z, Pinto M, Kros J, Oenema O (2014) The impact of the Nitrates Directive on nitrogen emissions from agriculture in the EU-27 during 2000–2008. Sci Total Environ 468–469:1225–1233

Wassmann P, Olli K (eds) (2006) Drainage basin nutrient inputs and eutrophication: an integrated approach. University of Tromsø. Available from http://kodu.ut.ee/~olli/eutr

Wulff F, Humborg C, Andersen HE, Blicher-Mathiesen G, Czajkowski M, Elofsson K, Fonnesbech-Wulff A, Hasler B, Hong B, Jansons V, Mörth CM, Smart JC, Smedberg E, Stålnacke P, Swaney DP, Thodsen H, Was A, Zylicz T (2014) Reduction of Baltic Sea nutrient inputs and allocation of abatement costs within the Baltic Sea catchment. AMBIO 43:11–25

Zillén L, Conley DJ, Andrén T, Andrén E, Björck S (2008) Past occurrences of hypoxia in the Baltic Sea and the role of climate variability, environmental change and human impact. Earth-Sci Rev 91:77–92

Chapter 3
Fisheries: A Case Study of Baltic Sea Environmental Governance

Piet Sellke, Marion Dreyer, and Sebastian Linke

Abstract This chapter analyses environmental governance through a case study of fisheries management in the Baltic Sea and investigates the problems, challenges and opportunities for improving sustainability in this sector. Fisheries management in the Baltic Sea is politically and culturally complex, institutionally fragmented and confronted with serious environmental problems, such as recent shifts in cod stocks. The central challenge is therefore to establish a regionally based, ecologically sustainable and socio-economically viable fisheries governance system for the Baltic Sea. Our analysis is focused on how past and current reform processes of fisheries management in the Baltic Sea have been able to move away from the path-dependent and highly ineffective management system linked to EU's *Common Fisheries Policy* towards new regional arrangements and procedures that address environmental problems in the Baltic on par with the social and economic challenges. We first describe existing governance structures for fisheries management in the Baltic Sea and their role in procedures of knowledge production, policy advice and decision-making. We then examine how the different governance actors (i.e. scientists, stakeholders, policymakers) address key issues such as the framing of the 'overfishing problem', the handling of uncertainty in the interactions of risk assessment and risk management and the role of stakeholder participation and communication. The chapter concludes by emphasising the need for an improved understanding of how scientific developments and connected uncertainty problems, policy constraints and stakeholder perspectives can be brought together for improving the biological, ecological and socio-economic sustainability of Baltic Sea fisheries governance.

Keywords EU fisheries management • Uncertainty • Stakeholder participation • Communication • Regionalisation

---

P. Sellke (✉) • M. Dreyer
DIALOGIK, Lerchenstrasse 22, 70176 Stuttgart, Germany
e-mail: sellke@dialogik-expert.de; dreyer@dialogik-expert.de

S. Linke
Department of Philosophy, Linguistics and Theory of Science, University of Gothenburg, Box 200, 405 30 Göteborg, Sweden
e-mail: sebastian.linke@gu.se

M. Gilek et al. (eds.), *Environmental Governance of the Baltic Sea*, MARE Publication Series 10, DOI 10.1007/978-3-319-27006-7_3

3.1 Introduction

One common resource that the nine countries bordering the Baltic Sea share is fish. The case study presented here deals explicitly with the 'risk of overfishing', which we define as the potentially adverse effects of fishing activities in terms of over-exploitation and subsequent unsustainable development of these activities from an economic, ecological and social point of view. The acute problems of over-exploitation in fisheries and their deteriorating effects on ecosystems and biodiversity as well as the resulting social and economic crisis have been addressed globally (e.g. McGoodwin 1990; Worm et al. 2009). In Europe these management failures are generally seen as a consequence of a nonworking governance framework for EU fisheries management (CEC 2009b[1]; Villasante 2011) which also holds true for the Baltic Sea (cf. Aps and Lassen 2010).

The main commercially exploited species in the Baltic Sea are cod, sprat, herring and salmon. Under the *Common Fisheries Policy* (CFP), which is the governing framework for fisheries management in the European Union, the main goal is to reduce the risk of overfishing and achieve a long-term sustainable exploitation of fish stocks. The CFP's objective is to maintain or restore fish stocks to levels that can produce the so-called maximum sustainable yield (MSY), which implies catching the largest possible proportion of a fish stock over an indefinite time period (Salomon et al. 2014). However, the use of the MSY concept in fisheries has been critiqued for not being sufficiently precise, both in terms of its conceptualisation in literature (e.g. Punt and Smith 2001) and its application in the daily practice of EU fisheries management (personal observations). When studying overfishing in the Baltic Sea, we cannot look at specific species in isolation, as the dynamics between different fish stocks (e.g. cod, sprat and herring) as prey–predator relationships form complex food–web relationships (Österblom et al. 2007).

One of the key challenges that fisheries governance in the Baltic Sea and elsewhere currently faces is a transition towards an *ecosystem approach to fisheries management* (EAFM), which takes multispecies considerations into account such as the implications of an increased cod stock on other fish stocks. The shift to such a management approach requires a substantive change in terms of the advice production system for fisheries management, which so far has mainly responded to the requirement of setting catch limits as so-called total allowable catches (TACs) and distributing fishing quotas amongst the Member States according to the principle of 'relative stability'. A crucial component of this TAC management system is an annual single-species approach (rather than a multispecies approach) which is difficult to reconcile with EAFM and, also, at a less complex level, with long-term, mixed fisheries management. The existence of this and other institutional impediments to the shift from an annual single fish stock management approach resulting from the TAC system to EAFM has been described as a situation of 'institutional

[1] Relevant publications before and after this are CEC (2001a, b, 2006, 2007, 2008, 2009b).

inertia' (Wilson 2009: 93). This institutional inertia is the central concern of this chapter.

Besides the problems with the TAC management approach, our case study also deals with the shift towards *results-based fisheries management* and the role of marine regions, Member States and stakeholders in such outcome-oriented management systems. Results-based management has been intensively discussed in the context of the recent reform of CFP, finalised in 2013. The European Commission's 2009 Green Paper on CFP reform suggested an approach where strategic decisions on principles and standards should remain at Community level (CEC 2009b), whilst decisions relating to technical implementation be delegated to Member States ideally organising themselves at the level of *marine regions* such as the Baltic Sea region (CEC 2009a). However, it is still unclear how the concepts of results-based management and a shifting burden of proof will be implemented through a regional approach under the new post-2013 CFP (Linke and Jentoft 2013; Nielsen et al. 2015).

In Wilson's publication *The Paradoxes of Transparency, Science and the Ecosystem Approach to Fisheries Management in Europe*, he highlights the importance of changing future fisheries management (Wilson 2009). Results-based management is identified as a promising alternative to current regulatory processes, which are overcentralised and top-down: 'The CFP can in many ways be argued to take the form of a classical intergovernmentalist, state-centric command-and-control, top-down management system' (Hegland 2009: 8). The European Commission's Green Paper (CEC 2009b) also recognised that the current CFP takes a top-down approach and needs to give the fishing industry more incentives to behave responsibly. In Wilson's and other's views, a 'nested results-based system, organised around both sets of economic activities and geographical areas' (Wilson 2009: 276; cf. Raakjaer and Tatenhove 2014), is better suited for governing an unpredictable and complex system such as the Baltic Sea's fisheries and essential for the implementation of EAFM. The role of stakeholders in CFP over the last 15 years is the other important topic taken up in this chapter. In particular, we look at how two main structures for stakeholder interaction within CFP – the Advisory Committee on Fisheries and Aquaculture (ACFA), set up in 1971 and now replaced by several specialised advisory councils, and the more recent Regional Advisory Councils (RACs) – inform CFP. The key research questions dealt with in this chapter with respect to the governance changes described here are: How is uncertainty dealt with in the governance interactions between risk assessment and risk management (science and policy)? And how does communication between different actors address differences in the framing of the central issues in today's fisheries governance structures? These fundamental governance issues are analysed and discussed. Our case study does not however deal with how external factors such as climate change, eutrophication (e.g. hypoxic bottoms), environmental pollution, invasive species or spreading diseases might contribute to decreasing fish stocks. Although these external pressures do exist, we do not treat them as part of the 'risk of overfishing'.

3.1.1 Research Approach and Analytical Framework

The case study on which this chapter is based was part of the RISKGOV project carried out from 2009 to 2011. The research team conducted 15 qualitative semi-structured interviews between February and October 2010 with stakeholders and actors involved in EU and Baltic Sea fisheries management. Six of these actors were from the policy sector (EU and country specific), two from science, three from industry, three from non-governmental organisations (NGOs) and one from a supra-national organisation (DG MARE). The research informing the analyses of this chapter is further complemented by findings from two more recent research projects undertaken between 2011 and 2015 as well as extensive document analyses and interviews with other key actors, particularly from the Baltic Sea RAC. Finally, it also involved participatory observations of meetings connected to the implementation of the 2013 CFP reform and the new regional Baltic Member States forum BALTFISH. However, the concrete effects of this most recent CFP reform have not been investigated empirically and are only partially integrated in this study.

Six main aspects have been especially relevant for this case study and serve as the analytical framework for our analyses:

- *Framing*: As with many decision-making processes in the area of risks and technology, the starting point of a governance process is making the implicit definitions of the current situation and of the current problem transparent. Different stakeholders often ascribe different meanings to the same term. Making these meanings explicit is referred to as 'framing' in the governance literature (cf. Renn 2008). In the context of this case study, we were interested in whether and how stakeholders have divergent views about the problem of overfishing and whether the term 'overfishing' is based on a shared understanding by different stakeholders.
- *Governance structures*: The qualitative analysis and interviews with policy experts, scientists, industry and non-governmental stakeholders aimed at detecting existing governance structures in the area, both overt and covert ones. This part of the study dealt explicitly with institutional issues and questions arising out of the debate on regionalisation as an asset or substitute to current governance structures.
- *Interaction of science and management*: Fisheries' and maritime management in general are dependent on knowledge about fish stocks and their interactions. Assessing fish stocks is a complex scientific task because cause–effect relationships are influenced by many intervening variables. For example, the development of a certain fish stock is not only related to fishing capacities and limits but also to the specific species' prey or predator (e.g. seals), maritime pollution, parasites and many other variables such as salinity, all of which are important in the Baltic Sea ecosystem. Within the EU fisheries governance structure, the task of resolving natural complexity is left to the scientists who give advice to management and decision-making bodies. The interaction of science and management is therefore of fundamental importance. How scientific advice is transferred

to the management sphere and how the scientific sphere takes up demands from management are two crucial aspects to enable a sound governance process. A particular focus in this study is the shift to an ecosystem-based approach to management and resolving uncertainty in the data.

- *Different kinds of incomplete knowledge in science and management*: The knowledge about a risk can be classified in three distinct categories (cf. Sellke and Renn 2010): complexity, uncertainty and ambiguity. Complexity is the difficulty of establishing a cause–effect relationship, for example, due to intervening variables or the multiplicity of variables included in the assessment. Complexity is, therefore, a cognitive conflict, i.e. there is not enough knowledge on all influencing variables and thus more knowledge has to be generated. Uncertainty often results from unresolved complexity but is also an entity of its own. It suggests that we do not know what we do not know. The lack of knowledge about unexpected and unknown effects can be due to stochastic relationships, ignorance or system boundaries. In the case of uncertainty, an evaluative conflict is posed. The third category, ambiguity, refers to the fact that risk assessment results can be accepted by different stakeholders as being sound science and valid but can also be interpreted differently in terms of the hazardous effects the risk might have. Further, normative, religious or other ethical considerations might overrun a technical risk assessment, for example, if for some stakeholders, only one of the common three pillars of sustainability is of value. Ambiguity thus stands for normative conflicts. The effects of the different kinds of uncertainty in science and management and how they are dealt with in Baltic fisheries are addressed here.
- *Communication with and amongst stakeholders*: Previous research in different risk-related areas has shown that misunderstanding with regard to the framing of an issue often leads to a flawed communication process later on (cf. Renn 2008). Further, in risk governance processes, specific forms of communication are appropriate at a given time; thus, not all communication (and participation) efforts are suitable to all occasions. How actors communicate with each other, however, is of broader and more fundamental importance to the whole governance process. Besides formal communication structures between one institution and another, informal communication channels between stakeholders are of equal importance. The changing role of stakeholders' involvement in Baltic Sea fisheries governance is therefore discussed under this heading.
- *Improvement opportunities for dealing with fishing*: Which measures, structures and changes in the management and/or advice system are of importance to the interviewee from his/her specific angle? What recommendations could be developed from the views articulated? From the specific point of view of the respective interviewee, certain insights might be found that cannot be seen from an outsider's perspective. The research process is, therefore, open to ascertaining the interviewees' thoughts and experiences with regard to improving management.

The six aspects that we focus on are partly overlapping and are dealt with in this paper in the following manner. First we describe the governance structures of EU fisheries management with CFP (Sect. 3.2). This is relevant background information for the results section (Sect. 3.3). The section addresses the two key aspects of assessment–management interactions and stakeholder communication processes. We present the results of our analysis of the interactions between risk assessment and risk management (i.e. how data is collected, communicated and fed into the policy process). We also describe the results concerning communication with and amongst stakeholders including role framing, transparency and discourse play in the communication processes. The presentation of the results is followed by a discussion section (Sect. 3.4); the paper concludes with some thoughts and recommendations (Sect. 3.5).

3.2 Governance Structures

3.2.1 The EU's Common Fisheries Policy

Fisheries management is one of the few areas where Member States have given EU institutions full decision-making power.[2] The exclusive right of the EU to manage fisheries is set forth in the Treaty of Lisbon (Article 2B), which states that the EU will have 'exclusive competence' over 'the conservation of marine biological resources under the common fisheries policy'.[3]

After the enlargement of the EU in 2004, all Baltic coastal states except Russia are now members of the EU. Since then CFP, originally established in 1983, is regulating the EU's fisheries activities. Prior to 2004, the states bordering the Baltic Sea managed internationally relevant issues of Baltic Sea fisheries multilaterally via the International Baltic Sea Fisheries Commission (IBSFC).[4] However, with EU enlargement IBSFC became redundant and ceased its activities on 31 December 2005. The two remaining parties, namely, the EU and Russia, arrange bilateral fisheries agreements, which are approved on behalf of the Community by Council Regulation (EC) No 439/2009 in 2009.

[2] Important regulations for the fisheries sector are EC No 2371/2002, EC 2004/585, EC 2187/2005, EC 1098/2007, EC 439/2009 and EC 1226/2009.

[3] European Union, Treaty of Lisbon Amending the Treaty on European Union and the Treaty Establishing the European Community, 13 December 2007, 2007/C 306/01, available at http://www.refworld.org/docid/476258d32.html [accessed 7 July 2014].

[4] The IBSFC was established in 1974 on the basis of the Convention on Fishing and Conservation of Living Resources in the Baltic Sea and the Belts (the Gdansk Convention) signed by the Baltic countries in 1973.

3.2.1.1 The TAC System

So-called total allowable catches (TACs)[5] are the key management measure for fisheries under CFP. These TACs are allocated amongst the EU Member States as 'national fishing quotas' on the basis of the 'relative stability' principle (Symes 1997), which ensures Member States a fixed percentage share of fishing opportunities for commercial species by taking into account countries' historical catch records before joining CFP.

TACs and quotas[6] are annually allocated for all commercially important fish stocks of the Baltic Sea. The EU Council of Ministers takes final decisions regarding TACs and related measures. The Council receives a proposal from the EU Commission for the following year's TACs and the conditions under which they should be caught.

The EU Commission is informed by *scientific advice* from the International Council for the Exploration of the Sea (ICES) in order to prepare proposals for TACs and other regulations (cf. Hegland 2009). ICES is an intergovernmental scientific organisation founded in 1902 and brings together more than 1,600 marine scientists from 20 countries to coordinate and promote marine science and provide scientific advice to a set of clients in response to their requests (see www.ices.dk). The biggest client of ICES is the European Commission with most advice requests coming from DG MARE and some from DG Environment. The OSPAR and HELCOM Conventions are also amongst ICES's regular clients. With its permanent secretariat in Copenhagen, Denmark, the main part of ICES's work is carried out by more than 100 working groups, assigned to specific topics of research. Whilst analysis of the conditions of fish stocks remains the main job of ICES in the fisheries area, the scientific organisation tries to increasingly provide advice at ecosystem level to support the intended shift towards a more holistic approach to managing Europe's seas (cf. Stange et al. 2012).

Scientific advice from ICES is also received by and channelled through the European Commission's own Scientific, Technical and Economic Committee for Fisheries (STECF) which is made up of scientists and experts, particularly in the fields of marine biology, marine ecology, fisheries science, fishing gear technology and fishery economics. It is the task of STECF to review the assessment of ICES with regard to biological, ecological, technological and economic issues and to pass it on to the Commission's Directorate-General for Maritime Affairs and Fisheries (DG MARE).

DG MARE, in addition to obtaining the scientific advice from ICES and STECF's evaluation of it, also executes a consultation process, which includes receiving advice from two *stakeholder structures*, the Advisory Committee on Fisheries and

[5]Under EU law, TAC means the quantity that can be taken from a stock each year. After CFP reform in 2013 and its ban to discard fish at sea, TACs now refer to the actual catch rather than landings.

[6]Under EU law, 'quota' means a proportion of the TAC allocated to the Community, a Member State or a third country.

Aquaculture (ACFA), which provides advice from industry to the Commission on fisheries issues, and the Regional Advisory Councils (RAC), created after the 2002 CFP reform and finally established between 2004 and 2008.[7] Economic interests have been predominant in ACFA's structure, even more important than economic issues are to RACs[8] (Wilson 2009: 96). ACFA, unlike RACs, mainly represents industry organisations at the European level.

The output of DG MARE is in the form of proposals to the EU Council of Ministers for decision-making. The most important of these decisions relate to setting the annual TACs for each species. Quotas of Member States, decided by the principle of 'relative stability', are thereupon distributed by national governments to their own operators who fish in national waters. However, although European Member States have given decision-making power to the EU, according to the principle of subsidiarity, they also carry responsibilities. Since the Treaty of Lisbon was signed in 2007, decision-making is shared between the EU Council of Ministers and the European Parliament although decisions on TACs still fall outside their joint purview.

3.2.1.2 Towards a Regionalised Results-Based Management?

Based on the principle of subsidiarity, it is up to the Member States as to how they distribute their quota allocation to their fishermen. It is also the responsibility of the Member States to pass the relevant laws and regulations and monitor and enforce compliance.

Under the pre-2013 CFP, there were a number of initiatives taken and *elements for a regionalisation* of the governance of risks pertaining to unsustainable fisheries:

– The CFP recognises the specificity of Europe's different seas and oceans by grouping technical measures into *regional regulations*. This included Council Regulation (EC) No 2187/2005 of 21 December 2005 for the conservation of fishery resources through technical measures in the Baltic Sea, the Belts and the Sound.

[7] Seven Regional Advisory Councils have been created since 2004 under CFP. Five are based on geographically and biologically coherent zones. Besides the Baltic Sea, other zones include the Mediterranean Sea, the North Sea, the Northwestern Waters and the Southwestern Waters. The two other RACs are based on the exploitation of certain stocks: pelagic stocks in Community waters (except in the Baltic and Mediterranean Seas) and high sea fisheries outside Community waters. With CFP reform in 2013, the name changed to merely Advisory Councils (AC), and three additional ones were set up for Aquaculture, Black Sea and distant fisheries (Hatchard and Gray 2014; Linke et al. 2014; Long 2010).

[8] The fisheries sector has been assigned a predominant position within RACs: in both the General Assembly and the Executive Committee, a majority (two thirds) of the seats are allotted to representatives of the fisheries sector and only one third to representatives of other interest groups (EC 2004, Art. 5(3)). This distribution of seats was changed in the 2013 CFP reform to a 60:40 representation ratio.

- Under the 2002 CFP reform, a system of *Regional Advisory Councils* was established that included the Baltic Sea RAC. The RACs are a mainstay of the EU's revised CFP. They are meant to satisfy stakeholders' demands for better involvement and thereby reduce deficits in regulatory compliance and enforcement and enrich the decision-making process in the fisheries through the prioritising of stakeholders' knowledge and experience (Linke et al. 2011). Whilst the RACs mainly respond to EU and national policy proposals, they may also act on their own initiative by proposing ways for dealing with problems, which they feel need to be addressed. The Baltic Sea RAC, for instance, took initiative on the issue of deficits of control and enforcement in the Baltic cod fishery, by coming up with long-term management plans. It also convened a major conference in Copenhagen on control and compliance in the Baltic Sea in March 2007 at which a set of conclusions were drawn on how this major problem of deficits of control could be resolved (CEC 2009b).

In the wake of the 2013 reform of CFP, the regionalisation of fisheries management in the EU became more relevant and concrete (cf. Raakjær and Hegland 2012; Symes 2012), particularly in the case of the Baltic (Hegland et al. 2015). This regionalisation happened not only in fisheries management but also in other domains of EU marine governance (cf. Gilek and Kern 2015). Whilst CFP remains in charge of fisheries in the Baltic Sea, in 2009 a new council was formed amongst the region's Member States, the *Baltic Sea Fisheries Forum* (BALTFISH). BALTFISH was formalised in 2013 through a Memorandum of Understanding (MoU) between the eight EU Member States in the region (BALTFISH 2013) so as to enable cooperation on Baltic Sea fisheries management. It is empowered through delegated or implemented acts from the EU Commission (Council and Parliament 2013). As stated in Article 18(2) of the basic regulation, Member States of the region 'shall cooperate with one another in formulating joint recommendations' and 'consult the Advisory Councils' (ACs, the former RACs), whilst the Commission facilitates the possibility of other relevant scientific bodies contributing (ibid., 38). Due to the innovative and proactive approach established with BALTFISH, the Baltic fisheries context has been held up as a forerunner and role model by the previous Commissioner for Maritime Affairs and Fisheries, Maria Damanaki, for how regionalisation of fisheries management could be implemented across Europe.

3.3 Insights into Risk Assessment–Risk Management Interaction and Stakeholder Communication Processes

This section presents the results of our case study primarily based on qualitative interviews carried out with different actors in the context of Baltic Sea fisheries. First, Sect. 3.3.1 summarises the results of the interaction between the spheres of risk assessment (science) and risk management (policy). Thereafter, we present in Sect. 3.3.2 results on communication with and amongst stakeholders.

3.3.1 Interactions between Risk Assessment and Risk Management

Interactions between the spheres of risk assessment and risk management are crucial for any successful risk governance process. Results from risk assessment have to be translated and transformed into manageable actions and decisions, a process that is often complicated by the complexity and uncertainty of the subject matter. A successful link between risk assessment and risk management is often based on transparent communication and open dialogue. Conversely, unsatisfactory risk management results are often due to communication and interaction problems between the two spheres. The following sections analyse the interaction processes between those two spheres in the context of fisheries' management in the Baltic Sea.

In general, our interviews addressed questions pertaining to decision-making structure for fisheries' management and, closely related to that, regionalisation. Regionalisation refers to changes in CFP's decision-making structure. Divergent views exist as to whether regional bodies should work merely consultatively or if they should also receive decision-making powers. These issues are important given the context of the new CFP reform process and its implementation (especially regarding the discard ban), which occurs at the regional level of the Baltic Sea through cooperation with and via BALTFISH. The new CFP reform process puts ACs potentially in a more responsible role and encourages them to contribute more proactively to management plans and increase coordination and collaboration with the Member States in the specific region. A failure of ACs and Member States to do so would result in the EU reverting back to traditional top-down management processes. Increased coordination would serve as an incentive for RACs and Member States to intensify their roles as mentioned, for example, at a Commission seminar on implementation of the CFP reform (http://ec.europa.eu/fisheries/news_and_events/events/20131025/index_en.htm; cf. Linke and Bruckmeier 2015). The Baltic Sea is one area where CFP reform could be implemented because a cooperative Member State forum, namely, BALTFISH, already exists, which could facilitate integration and harmonisation of stakeholders.

All of our interviewed actors, regardless of their institutional affiliation, supported a stronger emphasis on regionalisation. However, different views existed on how far regionalisation should go, how it can and should be implemented and what the real objectives of this regionalisation process should be. In discussing the prospects for regionalising CFP, Symes (2012) fears the EU would be in legal limbo and hence the regionalisation process would end with 'a note of frustration', i.e. 'that regionalising the CFP will be decided on legal and procedural grounds rather than from a perspective of good governance and what is best for the fisheries' (Symes 2012, 19).

None of the interviewees from science, industry, policy-making or the NGO sectors wanted to transfer decision-making power to the regional level alone. Although many of the actors would like to see *more* decision-making power transferred to the regional level, the lack of any involvement of institutions at EU level was seen to be more disadvantageous than advantageous. The need for an overarching structure for a region that is by nature multinational and comprises an open resource was acknowledged by all actors.

The following five subsections describe the process and challenges of generating the necessary knowledge to be used for decision-making processes. Challenges arise at each step, beginning with data collection through analysis and management of uncertainty.

3.3.1.1 Data Collection: The Analysis and Advice Process

The starting point of giving good scientific advice to the policy sector is the data collection process. Good data collection depends on the robustness of data collection procedures. National Fisheries Institutes (NFIs) collect data through survey vessels with high costs attached to the process (Wilson 2009: 96). Besides fisheries-independent data, fisheries-dependent data are also gathered, mainly through sampling of landings.

Inconsistencies between data collection across countries result in controversies about data gathered. This has been especially true with regard to discards of fish in relation to stocks. The heterogeneity of national authorities responsible for enforcement and inspection, as well as different scientific methodologies in relation to calculating the length or ageing of landings, also complicates comparisons. As Wilson (2009) describes, data from fisheries is often not regarded as reliable and conflicts around data sources seen to be politically laden. An example of the problems with unreliability highlights this. In 2004 it was estimated that landings of cod were under-reported by 35 % (Wilson 2009: 99). However, scientists were heavily dependent on this data. Consequently, unreliable data have negative effects on ICES' analyses later on in the process which could result in unreliable assessments of fish stocks and then possibly poorly formulated recommendations.

The system of data gathering for fisheries has improved through means like video surveillance on trawlers and sophisticated enforcement procedures that help especially with discard data. This applies particularly in relation to the discard ban, which was first implemented in January 2015 in the Baltic Sea.

The advice process has a very formalised procedure. First, the client asks for advice. Then expert groups (coordinated by ICES) have to collect data which is used to draft a technical scientific report. The draft report is then peer-reviewed by independent experts,[9] and the review as well as the draft report is used by the advice drafting group to give advice. This final 'ICES advice' is then agreed upon in the Advisory Committee (ACOM).

[9] This official external peer-review process was changed to an internal review in 2014.

3.3.1.2 Mitigation of a Basic Conflict Through Long-Term Planning

In 2002, the EU's CFP underwent a substantial reform process aimed to ensure sustainable exploitation of living aquatic resources. This was a landmark shift in EU fisheries governance. Since then, the conceptual basis for fisheries management under CFP is the maximum sustainable yield (MSY) principle, the precautionary approach and the ecosystem approach to fisheries management (EAFM). Key elements of this reform include (EC 2002):

- The adherence to the *precautionary approach* to protect and conserve living aquatic resources and to minimise the impact of fishing activities on marine ecosystems
- Adoption of a more *long-term approach* to fisheries management involving the establishment of multi-annual recovery plans for stocks outside safe biological limits and of multi-annual management plans for others stocks
- *Reduction of fishing effort* as a fundamental tool in fisheries management, notably in the context of multi-annual recovery plans
- Aim to progressively implement an *ecosystem-based approach* to fisheries management
- Increased involvement of the fisheries industry and other groups affected by CFP through the creation of *stakeholder-led Regional Advisory Councils*

Interviewees suggested that the move to a *long-term approach* to fisheries management has mitigated (or has potential to mitigate) conflict around TAC levels between scientists and conservation groups on the one hand and fishers and managers on the other hand. This conflict arose because management decisions in the past usually were only 'moderately responsive to ICES advice in setting TACs' (Patterson and Résimont 2007). Generally, ICES advice is an answer to the question: 'How much fish can we take this year without running the risk of not having enough left over for long-term exploitation?' (cf. Wilson 2009: 10). Over the last two decades, as Wilson argues, the TAC finally decided upon by the Council of Ministers is close to but not as much as ICES has advised. According to scientists, conservation groups and also Commission staff interviewed, the TACs decided upon are not based on carefully weighted biological and social and economic considerations. Instead, they claim that the divergence is due to putting short-term economic and social interests before long-term ecological imperatives. Short-term decision-making along with poor enforcement is seen as the main cause for increased stock depletion risk and economic risks for fisheries and fishers, an issue exhaustively discussed in the scientific literature (cf. Aps and Lassen 2010; Villassante et al. 2011). Over the last 15 years, despite the growing importance of sustainable development, it is believed by many including the media that scientific advice has largely been ignored in European fisheries (Wilson 2009: 28). In recent years, however, there is some indication that TAC decisions by the Council of Ministers have been more in line with scientific advice. As a result, the percentage of European stocks considered as overfished has declined (Lassen 2009, p. 6; EC 2012).

3.3.1.3 New Challenges to Scientific Advice with an EAFM

Long-term management approaches and EAFM in particular present huge, new challenges for scientific advice. Moving to these new forms of more holistic, ecosystem-based management will require substantial changes to processes within the scientific advisory system. This is currently high on the agenda in ICES and addressed, for example, through various new ecosystem working groups (e.g. WGIAB 2014; WKRISCO 2014). However, from the point of view of the managers in the European Commission, a historic *gap* has emerged that impedes management today, namely, between the form of advice that the scientific system is geared to provide and the form of advice that is progressively required under the revised CFP and EAFM context (Wilson 2009: 120). The shift towards long-term management is accompanied by a move towards management based on fleets and fisheries rather than single fish stocks. The move towards an EAFM is accompanied by a shift towards management of multispecies rather than single species.[10] Notwithstanding these shifts, there are still many different views about what an EAFM approach should be about. EAFM is often referred to as a fisheries management policy, which addresses issues such as by-catch of marine mammals and birds and the impact of fishing on the sea bottom. However, fisheries scientists find it challenging to provide advice because they are primarily trained to deal with fish stock units and examine single species one by one. In other words, they are used to deliver advice on TAC and its imperative of setting and distributing fishing quotas (Wilson and Delaney 2005).

An EAFM, on the other hand, is meant to capture stakeholders' perspectives, as it is broad and all encompassing. Within the governance literature, EAMF-type processes are often called 'paralysis by analysis' (Renn 2008) because its overly inclusive nature may lead to inertia. Although the governance side of EAFM deals only with outcomes of the risk assessment, there are still doubts about the feasibility of EAFM. Scientific models used by ICES currently involve only a minor number of variables, partially because of a lack of data and also because a large number of intervening variables with stochastic relationships, natural variations and changed human behaviour become impossible at some point to calculate. Furthermore, fundamental problems arise with the framing of objectives aimed at protecting the ecosystem. Should the sea be treated like a farm that aims to address long-term food production or should it rather be protected for its own sake (cf. Wilson 2009: 170)? As Wilson points out, based on attitude surveys amongst scientists working for ICES, divergent world views have severe effects on specific management measures like the precautionary approach. He also noted that there was a significant difference between agency and non-agency scientists (Wilson 2009: 171).

NGO representatives that we interviewed favoured EAFM as an approach because it is a holistic perspective and basically deals with all variables that are important for integrated *maritime* management. They see it as an approach that

[10] In 2015, a new multispecies management plan for cod, herring and sprat in the Baltic Sea is to be adopted by the EU Parliament as has already been done by the Council of Ministers.

overcomes the missing links between different biological, social and economic aspects concerning the Baltic Sea and that it improves communication. At the same time, they feel that more inclusive stakeholder processes should be launched and that regionalisation plays a crucial role in implementing EAFM. EAFM, however, also warrants the introduction of decision-making processes at the regional level. Some NGOs would like to see the inclusion of national decision makers into the current RAC structure.

Industry representatives interviewed in our study widely criticised EAFM, and if they did not, they saw it as a multispecies approach. Their criticism was based on the fact that EAFM was a tool by which different interest groups were able to get a voice on particular issues such as climate change or bird protection within the management system. Second, they said that the EAFM rather than being a tool for sectoral protection became a way to protect individual interests. Furthermore, even if EAFM was to be taken seriously, it would be impossible to implement because there were simply too many variables to include, many of which scientist have no knowledge about. In other words, EAFM is more of a utopian vision filled with flaws.

How the implementation of EAFM in EU fisheries and particularly in the Baltic Sea will develop is an issue left to further investigation. Recent scientific developments in ICES have been aimed at establishing so-called integrated ecosystem assessments (IEAs) as 'a formal synthesis tool to quantitatively analyse information on relevant natural and socio-economic factors, in relation to specified management objectives' (Möllmann et al. 2014). What remains to be seen is whether IEA development will succeed in becoming 'scientifically credible and socially legitimate' by integrating ecological, economic and social knowledge for marine governance in particular ecoregions (WKRISCO 2014).

Leading scientists from ICES that we interviewed have suggested that EAFM would be a significant challenge. In their view, it is next to impossible to connect multiple variables from different scientific disciplines and with different data methodologies into one model. The most that is possible in their view is a multispecies approach. However, regardless of their scepticism, ICES is preparing for EAFM through a working group.

Risk managers we interviewed had different views regarding EAFM. Some of them felt that EAFM will be the future of maritime management, whilst others shared the views of scientists. The differences in opinion largely stem from a different understanding of ecosystem-based management, i.e. whether it is aimed at the whole ecosystem or whether it targeted at a multispecies approach.

3.3.1.4 Uncertainty as a Key Challenge

Risk is a potential consequence (negative or positive) of human endeavour to obtain something they value (cf. Renn 2008). Risk assessment is the array of methods to assess hazards and vulnerability to these hazards. National experts working for ICES are in charge of assessing the vulnerability of the Baltic Sea and its fish stocks as well as the potential hazardous consequences of fishing in terms of discards. Risk assessment is always dependent on what we know about risk; in other words the task of risk assessment is to generate more knowledge about risk.

According to European Community legislation, CFP shall be guided by 'a decision-making process based on sound scientific advice which delivers timely results' (Council Regulation 2002, (Art. 2(2)b)). However, managers need to base their decisions on information which is associated with *considerable scientific uncertainty*[11]: There is uncertainty about how fish stocks will react to pressures, both human and environmental, and there is uncertainty involved in measuring existing fish stocks (due to sampling problems and misreporting of landings and discards) (Cochrane 2000; Finlayson 1994; Hawkins 2007). This type of uncertainty about size and age composition of fish stocks can result in incomplete knowledge that leads to ICES having difficulty making assessments. Consequently, giving advice to the Commission is also problematic. An example of this was the case of the Eastern Baltic cod stock in 2014 when an ICES assessment failed to provide adequate feedback on the present stock of cod (cf. Eero et al. 2015). The uncertainty challenge has therefore been and is still high on the agenda of both scientists and politicians, particularly because the TAC system in general has not been able to resolve the problems of overfishing and resource depletion (Lassen et al. 2014; Villasante et al. 2011).

CFP aims at responding to these perennial problems and growing insights vis-à-vis ecological issues and recognises the need to adopt a *precautionary approach* and, progressively, to move from a single-species-based fisheries management towards EAFM (EC 2002; Howarth 2008). As the interconnectedness between fisheries and the environment is still imperfectly understood, assessors and managers are faced with an even greater uncertainty challenge when ecosystem considerations are taken seriously: 'We have to accept that uncertainty in the science inputs to management will be larger (and more realistic) in an EAF…'[12] (Rice 2005: 269).

In summary, our case study interviews highlight that there is agreement amongst scientists and managers that EAFM requires an 'adaptive management' approach to deal with the complex and dynamic nature of ecosystems and the lack of full knowledge or understanding of how ecosystems function (cf. Linke and Bruckmeier 2015). The term 'adaptive management' is used to refer to a management approach that contains elements of learning-by-doing or research feedback, which make it possible to respond to such uncertainties (CEC 2008: 7f.).

3.3.1.5 Disagreement About Uncertainty Characterisation

ICES is in charge of generating knowledge about fish stocks for risk assessment. The process of how the necessary knowledge is created to assess further measures regarding fish stocks has been described above. Whilst ICES tries to gather the best available knowledge, issues of complexity cannot be addressed well within the

[11] Additionally, there is considerable complexity in defining cause–effect relationships.

[12] EAF means ecosystem-based approach to fisheries.

current system. Any scientific analysis is only as good as the data used, and data on fish stocks gathered by survey vessels and partially validated by fisheries' data is often unreliable and always incomplete. Further, ecosystems such as maritime environments are a dynamic entity with stochastic effects and uncertain dynamics.

There are increasing calls for improved concepts of uncertainty treatment and improved methodology in the characterisation, consideration, communication and management of uncertainty for scientific assessment and advice. It has been argued, for example, by Dankel et al. (2012), that a better understanding of how to characterise scientific uncertainty and its implications are needed. However, according to our interviewees, the underlying problem has still not been resolved, namely, that DG MARE wants numbers, whilst ICES's scientists prefer to give more qualitative and nuanced information.

This poses a fundamental problem for the whole management process: ICES is expected to deliver recommendations as sound and clear as possible in order for DG MARE to draft proposals for action. ICES' advice, on the other hand, has to be legitimate and definitive, i.e. it cannot be open to different interpretations by different stakeholders (Wilson 2009: 124). That there is uncertainty, however, has to be somehow communicated for the advice to become credible, but how exactly uncertainty should and can be communicated to DG MARE is an ongoing discussion, not least because of different understandings of uncertainty between ICES and DG MARE. This is referred to as 'institutional uncertainty' (Linke et al. 2014).

Industry representatives in all countries generally felt more comfortable with management decisions made by the Council of Ministers, partly because the latter did not fully follow the advice of ICES. Industrialists felt that NGOs read ICES' advice all too literally. They on the other hand understood the uncertainties involved and thus expected decision makers to set different TACs than those of ICES. Unsurprisingly, NGO representatives were very much in favour of the precautionary approach to scientific uncertainties. They felt that given the lack of or the unreliability of data, the precautionary approach would better address concerns of sustainable management of resources. Representatives from ICES on the other hand were less perturbed. They saw themselves as delivering a service to DG MARE based on the available resources and knowledge. In their view, their assessment was mainly a biological–economic one. They did not see a threat of species extinction. Rather, they were concerned that if TACs were set too high, economic problems might arise. DG MARE representatives saw the problem similarly and referred mainly to the challenges of data reliability and communication of uncertainty. Further, they felt that local and anecdotal knowledge about fish stocks should be taken seriously.

3.3.2 *Stakeholder Communication Processes*

Communication amongst stakeholders, but also between stakeholders and institutions managing maritime affairs in the Baltic Sea, is a crucial element of the whole governance process. Failed communication as a result of certain stakeholders being excluded or poor framing of issues can lead to the rejection of assessment results, advice, recommendations and/or management decisions. It is therefore important to consider both formal communication channels and informal ones. The former add legitimacy to governance processes, whilst the latter have the potential to build trust and cross-sector coalitions for collaboration in management (cf. Linke and Jentoft 2013).

Even before communication on a specific issue begins, the *framing of the issue* is critically important. Whether something is seen as an opportunity or a risk or a threat or a challenge or should be looked at in terms of its economic, ecological or social effects or at all of them simultaneously determines subsequent communication processes.

3.3.2.1 Framing: The Issue of 'Overfishing'

Results from the interviews carried out suggest that there are highly divergent meanings attached to the term 'overfishing' by the main actors involved in European fisheries governance which consequently result in conflict amongst the actors. Today the EU Commission uses the term 'overfished' whilst addressing concept of maximum sustainable yield (MSY). The 2009 Green Paper says:

> While a few EU fleets are profitable with no public support, most of Europe's fishing fleets are either running losses or returning low profits. Overall poor performance is due to chronic overcapacity of which overfishing is both a cause and a consequence: fleets have the power to fish much more than can safely be removed without jeopardising the future productivity of stocks. (CEC 2009a: 7)

The Fisheries Secretariat[13] states hereupon:

> In 2009 the Eastern cod stock was described as being overfished with respect to the potential long term yield for the fishery, meaning that the stock could be much larger and more fish could be caught in the future if fishing mortality was reduced.

The Green Paper's statement '88 % of Community stocks are being fished beyond MSY' therefore does not mean that these stocks are near to collapse but 'that these fish populations could increase and generate more economic output if they were left for only a few years under less fishing pressure' (CEC 2009a: 7). The concept 'outside safe biological limits' on the other hand refers to the more serious situation of overfishing, implying that these stocks 'may not be able to replenish' (ibid.).

Some interviewees said that conservation and environmental groups collapse the distinction between 'overfished stocks' and stocks *near to collapse* something that the media also does. According to many industry representatives and even other stakeholders, treating everything as overfished stocks is overdramatic and a case of

[13] Information from the website of The Fisheries Secretariat, which describes itself as 'a non-profit organisation dedicated to work towards more sustainable fisheries at an international level, with a focus on the European Union', http://www.fishsec.org/article.asp?CategoryID=1&ContextID=194

false information being fed to consumers and the wider public. If the resulting misconceptions were to influence consumer behaviour, this could have negative effects for the fishing and processing industries.

Official statements by the EU, of the type given above, can be used by different stakeholders to make different points. Whilst industry representatives we spoke to felt that the EU was *not* saying anything about biological extinction of species, NGO representatives felt that the EU was stressing that overfishing was taking place although they did not clearly define overfishing. A fundamental problem with regard to the framing of the issue of 'overfishing' is that it can be defined economically, socially or environmentally. Actors largely choose perspectives that best fit their own agendas. Our interviews revealed that differences in the framing of the issue of overfishing were not so much about whether species are close to extinction but rather about whether a long-term or short-term perspective with regard to overfishing should and could be addressed by EAFM.

As this analysis illustrates, framing does not only relate to what is understood by different actors about the issues at stake but also to what rules, procedures and conventions specifically mean in dealing with risk. EAFM is a remarkable example of this phenomenon since different actors point out different aspects of EAFM. It is understood by some actors as merely a multispecies approach, whereas others employ a more holistic view of the whole environmental system. These differences in the framing of the concept of EAFM need to be communicated within the governance process. If communication is poor, actors might not be talking about the same issue when they refer to EAFM. Further, if the framing differs, the interpretation of rules, procedures and conventions will differ as well.

3.3.2.2 Enhancing Transparency in the Scientific Advisory System

In the context of recent restructuring of ICES, participation has been extended by opening up meetings to 'observers' much more than in the past (cf. Stange et al. 2012). Since 2004, ICES has been inviting representatives from industry and environmental NGOs to attend meetings of the Advisory Committee on Fishery Management (ACOM) (Wilson 2009: 122), which has representatives from each of the 20 ICES member countries and meets every year in the spring and autumn. This 'transparency through observers' (ibid., p. 274) was a response to demands from both DG MARE and stakeholder groups.

In 2013, the Working Group on Maritime Systems (WGMARS) put forward their suggestions on how to shape a more transparent process that also integrates stakeholders in the scientific advice process. Specifically, the report emphasised the need to encompass stakeholders' research needs over the medium and long term, evaluate and propose best practices in stakeholder engagement in EU-funded projects and define terms of reference for an ongoing dialogue with stakeholders and scientists (ICES WGMARS Report 2013).

Generally, most of our interviewees valued increased transparency about procedures pertaining to generating scientific knowledge and advice and stressed the need for ICES to adapt to new societal circumstances. In recent years ICES has therefore undertaken substantial efforts to make its scientific advisory processes more transparent, reliable and digestible for non-scientists. For example, since 2013 ICES produces an 'up-to-date and easy-to-read digest of the official ICES advice', which is a popularised version of official communications that are more understandable to a wider audience. Distributed as an online paper (see, http://www.ices.dk/publications/our-publications/Pages/Popular-advice.aspx) through, for example, social media, this popular advice presents ICES' new orientation towards non-scientific audiences. Whilst such initiatives are generally lauded amongst stakeholders and the general public, managers, for example, from DG MARE, feel that there is a need to discuss how to combine highly technical communications of scientific procedures 'at the bench' with more understandable communiqués for outreach.

3.3.2.3 Risk Communication to the Public

Industry and management representatives as well as scientists from ICES confirmed to us that 'overfishing' is a hot topic for the media. That fish stocks will be extinct e.g. by 2048 (cf. Stokstad 2009) is a good headline, but for industry representatives and managers, such headlines strain the governance process because it potentially exaggerates things. Hence, communication to the public should be through a shared framing process in terms of how the issue is framed.

Overall our interviewees did not see a need for more specific public participation within the governance process. Communication with the public was regarded as important in terms of sharing information. Although all interviewees agreed that it is always good to have the public involved (because of the complexity of the governance process), they did not feel that the public should make recommendations about future policies. Some of the interviewees, especially from DG MARE, saw the European Parliament as well as national parliaments as being the democratically elected representatives of the public.

3.4 Discussion

Our case study about Baltic Sea fisheries points out several problematic issues pertaining to a good governance process for fisheries management. Some of those issues are generic to the field of fisheries management itself, whilst others relate to organisational shortcomings which can be improved upon. Before making recommendations for such improvements, two basic issues need to be emphasised to point out key problematic areas:

1. *Dealing with uncertainty* in assessment–management interactions
2. *Communication and stakeholder participation*

3.4.1 Dealing with Uncertainty in Assessment–Management Interactions

Estimates of fish stocks in the Baltic Sea, as well as elsewhere, have to deal with complex cause–effect relationships involving many intervening variables. The risk governance literature, which deals with complexities between cause and effect relations, calls for the best available involvement of experts in order to achieve prudent assessment results (Renn 2008). With regard to EU fisheries governance structures, ICES takes up this task by devoting all its efforts to involve the best available (natural) scientific expertise in order to advice decision makers in a way that minimises *cognitive* conflicts.

Uncertainty complicates complex governance arrangements of risk assessment–risk management interactions in fisheries. Uncertainty is a major issue in the science–policy interface of EU fisheries management under CFP (cf. Dankel et al. 2012) and, as our results revealed, exists in several ways:

(a) *Uncertainty in data gathering*
(b) *Uncertainty in data analysis*
(c) *Uncertainty impacts stemming from points (a) and (b) in framing, evaluation and management*

The fishery sector is characterised by the so-called second-order uncertainty (Renn 2008), i.e. a risk situation where circumstances might change in an unpredictable and unsystematic manner as in the case of fish stocks and the environmental system of which they are part. Second-order uncertainty is difficult to communicate, hence, our focus on it. In the case of biological assessments of fish stocks, it is helpful to make a distinction between *aleatory* and *epistemic* uncertainty (Renn 2008: 71). Aleatory uncertainty characterises randomness in samples, which means that only in the long run and with a large enough sample can the distribution of possible values be identified. Epistemic uncertainty on the other hand stems from a lack of knowledge of dynamics or phenomena within the field. Although extended data gathering and research might decrease both aleatory and epistemic uncertainties, with dynamic systems such as marine environments, uncertainty often prevails and can even increase as a consequence of further research as it happened in 2014 with the failure of ICES' Eastern Baltic cod assessment (Eero et al. 2015).

Distinguishing between the two types of uncertainty and designing communication processes that make the distinction between the two obvious to the ICES audience (the audience being broader than the clients alone) will increase transparency about different aspects of uncertainty and their role in assessment–management interactions. Such transparency about the type of uncertainty is especially needed as uncertainty is always interpretable, i.e. a subject of 'interpretative flexibility' (Meyer and Schulz-Schaeffer 2006), and therefore always a potential source of conflict in discussions amongst stakeholders in Baltic Sea fisheries. This is most notable in the Baltic RAC, where NGOs and fisheries representatives read different things from the scientific reports of ICES (cf. Linke et al. 2011, 2014).

Biological assessments of fish stocks, particularly when applying EAFM, also have to deal with epistemic uncertainty. The conflict around epistemic uncertainty and how to interpret results cannot only be solved by further scientific analysis, as it is not merely a cognitive but also an evaluative conflict. At present the situation in EU fisheries governance leaves this question somewhat open. It is not clear as to who will discuss the implications of evaluative judgements that have to be made. The ways in which social aspects of uncertainty are currently addressed may play a crucial and yet underrated role in the further framing of EU fisheries management discourse, for example, with respect to contributions of BS RAC to the Baltic salmon management plan (Linke and Jentoft 2014). Further, a second field of conflict exists in terms of the interaction of science and management with the TAC-based management system: policy and decision makers require numbers (quota advice) as input for further decision-making and communication, whereas the system of science or scientists would produce more qualitatively driven information. Uncertainty can also stem from unreliable and in some cases invalid data. Data analysis is obviously dependent on sound data gathering procedures. Unfortunately, data gathering is one of the problematic aspects of the process. This is partly due to logistical and practical restrictions and partly to differing interests of national NFIs and fisheries, who are mainly in charge of data gathering (Wilson 2009). The more unreliable the data gathering process is, the more unreliable data analysis and advice based on it. NFIs certainly have national interests and are closely connected to the national management and policy sector with Member States being part of the European Council. Hence, whilst data gathering methods have improved, they still account for a degree of uncertainty.

How the role of uncertainty in management and decision-making is conceptualised and dealt with has in turn also impacted the framing of the problem, risk evaluation, risk management and risk communication. Different framing perspectives in terms of what role uncertainty actually plays in risk management can lead to different interpretations of assessment results, which subsequently has consequences in terms of different risk evaluations and management strategies (e.g. applying the precautionary principle or not).

3.4.2 Communication and Stakeholder Participation

Communication amongst actors, and particularly with and amongst stakeholders, plays a central role in any governance process. Communication processes in EU fisheries management have improved significantly over the last two decades, particularly through RACs, increased interaction between scientists and stakeholders (e.g. in the form of observer status being given to RAC members at ICES meetings and vice versa), and management bodies improving communication. Further, participation in planning and decision-making processes has increased significantly in regional decision-making processes as, for example, with the discard ban, to be implemented in line with Article 15 in the new CFP reform (Council & Parliament

2013, 35). Both CFP reform and the new multispecies plan to be finalised in 2015 offer possibilities of the new interactions between the regional triangle of Member States (via BALTFISH), the stakeholder sector (via the BS RAC) and the EU Commission and hence offer both challenges and new opportunities for improved communication between these new management bodies.

The problematic aspects mentioned with respect to uncertainty treatment also come from inadequate communication. One example is the BS RAC: although it is a forum for opposing views to come together (i.e. NGOs and industry), communication is hindered by the uneven representation ratio of the two parties in the BS RAC and also poor material resources for working effectively on the questions at stake. Further, because of different normative assumptions (e.g. long-term versus short-term value perspectives), uncertainty is interpreted differently by different actors and often cannot be reconciled (Linke et al. 2011; cf. Renn 2008).

Finally, communication is central to the interaction between scientists and management. Communication needs to be a two-way process in which mutual trust is built between actors. The goal should be to assist stakeholders in understanding risk managers' decisions and the rationale of risk assessment results (Renn 2008) so as to enable them to make informed choices in relation to their interests and values (Johannesen and Lassen 2014). To achieve these objectives, risk communication is a task for professionals, something which is rarely understood. Trust can be built if there is a willingness to admit that uncertainty exists in risk assessment results. Trust will also lead to legitimacy (Dankel et al. 2012; Renn 2008). We have illustrated based on our interviews that communication processes in the Baltic Sea fisheries sector are often lacking because trust is missing. Whether the new structures that have emerged after the 2013 CFP reform will improve mutual trust remains to be seen.

3.5 Conclusions and Recommendations

We have suggested in this chapter that the main challenges of fisheries governance in the Baltic Sea relate to risk assessment–risk management interactions and their treatment, communication of different forms of uncertainty, harmonisation of the (interdisciplinary) knowledge base and the organisation of stakeholder participation to improve communication. Based on our case studies, we conclude that current governance structures are not yet capable of fully addressing the problems of scientific uncertainty, interpretations of this uncertainty and connected misunderstandings amongst the different actors in terms of reaching desired outcomes of sustainable fisheries in the Baltic Sea. Increased interactions amongst individual actors, for example, in RACs as well as between different management organisations (e.g. ICES, RAC and the EU Commission), as well as more developed institutional and procedural designs for stakeholder involvement in management and decision-making at the regional level, are urgently needed for improving environmental governance of the Baltic Sea.

EAFM, whilst placing further demands on fisheries science and management and their interactions, is an important step towards a more holistic, regionalised and stakeholder inclusive fisheries governance approach for the Baltic Sea that could help balance environmental and social dimensions. However, whilst the theory behind EAFM with its novel knowledge requirements with regard to integrated pressures, ecosystem impacts and societal concerns is rather well developed (cf. Gilek et al. 2015; McLeod and Leslie 2009), a coherent strategy for EAFM implementation is still lacking. Expectations rest with science, especially all of ICES' ecosystem working groups (e.g. the ICES/HELCOM Working Group on Integrated Assessments of the Baltic Sea, WGIAB), to put forward new tools and methods to build a knowledge base that can achieve integration that EAFM demands (cf. Möllmann et al. 2014).

Furthermore, the EU's CFP reform process 'opened up' EU fisheries governance in the Baltic Sea context to an extended regionalisation approach. Such an approach has been strengthened due to the new Member States collaborating via BALTFISH. Other reasons for this strengthening include increased incorporation of EAFM in the new CFP as well as integration and empowerment of the stakeholder sector due to a slightly enhanced role for the reformed RACs (ACs). Whilst stakeholders' input to the EU authorities via ACs still remains purely advisory, as it was during the previous CFP period, under the new basic regulation, both the Commission and the Member States are legally bound to respond with detailed reasons in cases where adopted measures or regulations diverge from the recommendations and suggestions that were received from the ACs (Council and Parliament 2013, Article 44, p. 47). However, the new provisions flagged under efforts of 'regionalisation' of the new (post-2013) CFP still remain relatively weak in EU's fisheries governance (Hatchard and Gray 2014; Salomon et al. 2014).

Overall, as Symes (2012, 1) states, the current CFP of the new regionalisation approach 'presently faces the most important challenge of its thirty year history'. The challenge is ensuring that the short-term management approach of annual fishing quotas is changed and a new perspective embracing more fundamental changes aimed at long-term viability and sustainability of the fisheries sector adopted. Such an approach should remain true to the overall European project (ibid.). Symes and other scholars, whilst discussing these new challenges, are pessimistic about whether the authorities in 'technocratic Brussels' (Salomon et al. 2014, 81) are willing to delegate powers to the regional levels such as the Baltic Sea.

However, despite such legal constraints, the Baltic Sea is seen as the closest prototype for regional cooperation under the post-2013 CFP (cf. Hegland et al. 2015). The future will tell if and how the new regionalisation project for fisheries governance in the Baltic Sea including BALTFISH will result in the implementation of EAFM and a stakeholder inclusive management approach.

In summary our recommendations for improving environmental governance of the Baltic Sea fisheries relate first of all to knowledge aspects, which are of paramount importance for CFP in general and hence also for the Baltic Sea context. Interdisciplinary knowledge and specifically social science research are the need of the hour (Linke and Jentoft 2014; cf. Urquardt et al. 2014) and therefore have to be better integrated into the risk governance cycle of fisheries management under CFP. An increased emphasis on social aspects promises to create the necessary trust between actors and also to take local knowledge into account. Furthermore, communication processes need to focus more on two-way channels of communication rather than only on providing information and educating people. Taking into account the social, cultural and economic needs of different stakeholders around the Baltic Sea will be crucial for communicating management results more appropriately to stakeholders. Ultimately, better results depend on better sharing of management responsibility amongst actors, both with respect to the aims of the newly reformed CPF and regional fisheries management in the Baltic Sea. The current CFP with its interest-based system of stakeholder representation (Linke and Jentoft 2014) and its legally bound centralisation of decision-making seems to leave little room for such visionary objectives of shifting the burden of proof and creating a more responsible, results-based management system (cf. Linke and Jentoft 2013; Nielsen et al. 2015). Any serious implementation of recommendations put forth above is still lacking in the Baltic fisheries context at the time of writing.

Acknowledgements The research was funded by the Foundation for Baltic and East European Studies and the European Community's Seventh Framework Programme (2007–2013) under grant agreement No 217246 made with the joint Baltic Sea research and development programme BONUS, as well as by the German Federal Ministry of Education and Research (BMBF) and the Swedish Research Council FORMAS. SL also acknowledges funding from the Swedish Research Council and Riksbankens Jubileumsfond. We wish to thank all these institutions for enabling this research. Two peer reviewers are also thanked for valuable comments on an earlier version of the chapter.

References

Aps R, Lassen H (2010) Recovery of depleted Baltic Sea fish stock: a review. ICES J Mar Sci 67:1856–1860

BALTFISH (2013) Memorandum of understanding: principles and working methods of the Baltic Sea Fisheries Forum (BALTFISH)

CEC (Commission of the European Communities) (2001a) Green paper on the future of the Common Fisheries Policy. COM(2001) 135 final, Brussels, 20 Mar 2001

CEC (Commission of the European Communities) (2001b) European governance. A white paper, COM 428 Final. Brussels 25 Jul 2001

CEC (Commission of the European Communities) (2006) Communication from the Commission to the Council and the European Parliament. Implementing sustainability in EU fisheries through maximum sustainable yield. COM(2006) 360 final. Brussels, 4 Jul 2006

CEC (Commission of the European Communities) (2007) Communication from the Commission to the Council. Fishing opportunities for 2008. Policy statement from the European Commission. COM(2007) 295 Final. Brussels, 6 June 2007

CEC (Commission of the European Communities) (2008) Communication from the Commission to the Council and the European Parliament. The role of the CFP in implementing an ecosystem approach to marine management. COM(2008) 187 Final. Brussels, 11 Apr 2008

CEC (Commission of the European Communities) (2009a) The Common Fisheries Policy. A user's guide. Office for Official Publications of the European Communities, Luxembourg

CEC (Commission of the European Communities) (2009b) Green paper. Reform of the Common Fisheries Policy. COM(2009) 163 final. Brussels, 22 April 2009

Cochrane KL (2000) Reconciling sustainability, economic efficiency and equity in fisheries: the one that got away? Fish Fish 1(1):3–21

Council & Parliament (2013) Regulation 1380/2013 of the European Parliament and of the Council of 11 December 2013 on the Common Fisheries Policy

Council Regulation (EC) No 1098/2007 of 18 September 2007 establishing a multiannual plan for the cod stocks in the Baltic Sea and the fisheries exploiting those stocks, amending Regulation (EEC) No 2847/93 and repealing Regulation (EC) No 779/97. OJ of the European Union L 248/1, 22.9.2007

Council Regulation (EC) No 1226/2009 of 20 November 2009 fixing the fishing opportunities and associated conditions for certain fish stocks and groups of fish stocks applicable in the Baltic Sea for 2010. OJ of the European Union, L 330/1, 16.12.2009

Council Regulation (EC) No 2002. No. 2371/2002 on the conservation and sustainable exploitation of fisheries resources under the Common Fisheries Policy

Council Regulation (EC) No 2004/585/EC of 19 July 2004 establishing Regional Advisory Councils under the Common Fisheries Policy

Council Regulation (EC) No 2187/2005 of 21 December 2005 for the conservation of fishery resources through technical measures in the Baltic Sea, the Belts and the Sound, amending Regulation (EC) No 1434/98 and repealing Regulation (EC) No 88/98. OJ of the European Union, L 349/1, 31.12.2005

Council Regulation (EC) No 439/2009 of 23 March 2009 concerning the conclusion of the agreement between the European Community and the Government of the Russian Federation on cooperation in fisheries and the conversation of the living marine resources in the Baltic Sea. OJ of the European Union, L 129/1, 28.5.2009

Dankel DJ, Aps R, Padda G, Röckmann C, Sluijs JP, Wilson D, Degnbol P (2012) Advice under uncertainty in the marine system. ICES J Mar Sci 69(1):3–7

Eero et al (2015) Eastern Baltic cod in distress: biological changes and challenges for stock assessments. ICES J Mar Sci. doi:10.1093/icesjms/fsv109

European Commission (EC) (2012) Communication from the Commission to the Council concerning a consultation on fishing opportunities for 2013. COM 278 final

Finlayson AC (1994) Fishing for truth. A sociological analysis of northern cod stock assessments from 1977–1990. Institute of Social and Economic Research of Memorial University of Newfoundland, Newfoundland

Gilek M, Kern K (eds) (2015) Governing Europe's marine environment. Europeanization of regional seas or regionalization of EU policies? Ashgate, Farnham

Gilek M, Karlsson M, Udovyk O, Linke S (2015) Science and policy in the governance of Europe's marine environment – the impact of Europeanization, regionalization and the ecosystem approach to management. In: Kern K, Gilek M (eds) Governing Europe's marine environment. Europeanization of regional seas or regionalization of EU policies? Ashgate, Farnham, pp 141–163

Hatchard J, Gray TS (2014) From RACs to advisory councils: lessons from North Sea discourse for the 2014 reform of the European Common Fisheries Policy. Mar Policy 47:87–93

Hawkins AD (2007) Review of science and stakeholder involvement in the production of advice on fisheries management. Work Package 4 paper of the EU project "Scientific Advice for Fisheries Management at Multiple Scales" (SAFMAMS). Available from: http://www.ifm.dk/safmams/Downloads/WP3/D2%20Review%20of%20Science%20and%20Stakeholder%20Involvement.pdf
Hegland TJ (2009) The Common Fisheries Policy and competing perspectives on integration, Publication Series Department of Development and Planning. Aalborg University, Aalborg
Hegland TJ (2012) Fishing for change in EU governance: excursions into the evolution of the Common Fisheries Policy. PhD thesis, Aalborg University, Aalborg
Hegland TJ, Ounanian K, Raakjer J (2012) Why and how to regionalise the Common Fisheries Policy. Marit Stud 11(7)
Hegland TJ, Raakjaer J, van Tatenhove J (2015) Implementing ecosystem-based marine management as a process of regionalisation: some lessons from the Baltic Sea. Ocean Coast Manag (In press)
Howarth W (2008) The interpretation of 'precaution' in the European Community Common Fisheries Policy. J Environ Law 20:213–244
ICES WGMARS Report (2013) Interim report of the Working Group on Maritim Systems (WGMARS), 4–8 Nov 2013, Stockholm, Sweden. ICES CM 2013/SSGSUE:06.30pp
Johannesen O, Lassen H (2014) Decision-making management procedures. Cost – efficiency – democracy in selected procedures in maritime spatial planning. TemaNord 2014:532
Lassen H (2009) Scientific advice for Fisheries Management. In Fisheries, sustainability and development. Royal Swedish Academy of Agriculture and Forestry, Stockholm
Lassen H, Kelly C, Sissenwine M (2014) ICES advisory framework 1977–2012: from Fmax to precautionary approach and beyond. ICES J Mar Sci 71:166–172
Linke S, Bruckmeier K (2015) Co-management in fisheries – experiences and changing approaches in Europe. Ocean Coast Manag 104:170–181
Linke S, Jentoft S (2013) A communicative turnaround: shifting the burden of proof in European fisheries governance. Mar Policy 38:337–345
Linke S, Jentoft S (2014) Exploring the phronetic dimension of stakeholders' knowledge in EU fisheries governance. Mar Policy 47:153–161
Linke S, Dreyer M, Sellke P (2011) The regional advisory councils. What is their potential to incorporate stakeholder knowledge into fisheries governance? AMBIO 40(2):133–144
Linke S, Gilek M, Karlsson M, Udovyk O (2014) Unravelling science-policy interactions in environmental risk governance of the Baltic Sea: comparing fisheries and eutrophication. J Risk Res 17(4):505–523
Long R (2010) The role of Regional Advisory Councils in the European Common Fisheries Policy: legal constraints and future options. Int J Mar Coast Law 25:289–346
Mc Goodwin JR (1990) Crisis in the world's fisheries: people, problems, and policies. Stanford California University Press, Stanford
McLeod KL, Leslie H (eds) (2009) Ecosystem-based management for the oceans. Island Press, Washington, DC
Meyer U, Schulz-Schaeffer I (2006) Three forms of interpretative flexibility. In: Science, technology & innovation studies, Special Issue 1
Möllmann C, Lindegren M, Bleckner T, Bergström L, Casini M, Diekmann R, Flinkman J, Müller-Karulis B, Neuenfeldt S, Schmidt JO, Tomczak M, Voss R, Gårdmark A (2014) Implementing ecosystem-based fisheries management: from single-species to integrated ecosystem assessment and advice for Baltic Sea fish stocks. ICES J Mar Sci 71(5):1187–1197. doi:10.1093/icesjms/fst123
Nielsen KN, Holm P, Aschan M (2015) Results based management in fisheries: delegating responsibility to resource users. Mar Policy 51:442–451
Österblom H, Hansson S, Larsson U, Hjerne O, Wulff F, Elmgren R, Folke C (2007) Human-induced trophic cascades and ecological regime shifts in the Baltic Sea. Ecosystems 10:877–889

Patterson K, Résimont M (2007) Change and stability in landings: the responses of fisheries to scientific advice and TACs. ICES J Mar Sci 64(4):714–717
Punt AE, Smith ADM (2001) The gospel of maximum sustainable yield in fisheries' management: birth, crucifixion and reincarnation. In: Conservation of exploited species, Cambridge University Press, Cambridge
Raakjær J, Hegland TJ (2012) Introduction. Regionalising the Common Fisheries Policy. Marit Stud 11:5
Raakjaer J, Tatenhove J (2014) Marine governance of European seas. Mar Policy 50(Part B):323–382
Renn O (2008) Risk governance: coping with uncertainty in a complex world. Earthscan, London
Rice JC (2005) Implementation of the ecosystem approach to fisheries management – asynchronous co-evolution at the interface between science and policy. Mar Ecol Prog Ser 300:265–270
Salomon M, Markus T, Dross M (2014) Masterstroke or paper tiger – the reform of the EU's Common Fisheries Policy. Mar Policy 47:76–84
Sellke P, Renn O (2010) Risk, society and environmental policy: risk governance in a complex world. In: Gross M, Heinrichs H (eds) Environmental sociology: European perspectives and interdisciplinary challenges. Springer, Dordrecht, pp 295–323
Stange K, Olsson P, Österblom H (2012) Managing organizational change in an international scientific network: a study of ICES reform processes. Mar Policy 36:681–688
Stokstad E (2009) Détente in the fisheries war. Science 324:170–171
Symes D (1997) The European community's common fisheries policy. Ocean Coast Manag 35(2–3):137–155
Symes D (2012) Regionalising the Common Fisheries Policy: context, content and controversies. Marit Stud 11(6)
Urquhart J, Acott T, Symes D, Zhao M (eds) (2014) Social issues in sustainable fisheries management. Springer, Dordrecht
Villasante, LS (2011) Relations between the economy, co-adaptability and resilience of marine ecosystems. Revista Galega de Economia, University of Santiago de Compostela. Faculty of Economics and Business, vol 20(2)
WGIAB (2014) Second interim report of the ICES/HELCOM Working Group on Integrated Assessments of the Baltic Sea. ICES CM 2014/SSGRSP:06
Wilson DC (2009) The paradoxes of transparency. Science and the ecosystem approach to fisheries management in Europe. Amsterdam University Press, Amsterdam
Wilson DC, Delaney AE (2005) Scientific knowledge and participation in the governance of fisheries in the North Sea. In: Gray T (ed) Participation in fisheries governance. Kluwer Academic Publishers, Dordrecht
WKRISCO (2014) Report to ICES ACOM/SCICOM Committee. ICES CM SSGBENCH:01.
Worm B, Hilborn R, Baum JA et al (2009) Rebuilding global fisheries. Science 325:578–585

Chapter 4
Biological Invasions: A Case Study of Baltic Sea Environmental Governance

Katarzyna Smolarz, Paulina Biskup, and Aleksandra Zgrundo

Abstract This chapter describes bioinvasions as an example of a relatively new crosscutting domain of environmental governance whose management is affected by a high level of uncertainty, complexity and ambiguity. In the Baltic Sea region, legislation and policies related to invasive alien species (IAS) are still under development, and as a consequence, there are a few legally binding instruments dealing with the problem. Due to the fact that environmental changes linked to other environmental risks (eutrophication, maritime transportation, climate change) may intensify biological incursions, development of a uniform policy, followed by its ratification among EU Member States in the Baltic Sea region as well as Russia, is generally seen as a top priority for many actors involved in environmental governance. Hence, the adoption of a precautionary approach and the Ecosystem Approach to Management (EAM) driven by precise goals and executed by policies and best practices are proposed as holistic and necessary management tools for preventing and controlling bioinvasions. This chapter focuses on barriers and opportunities for the implementation of the EAM concept and on identifying possible ways to improve the effectiveness of IAS management. Finally, we argue that biological invasions and in particular their consequences may impact on a wide set of ecosystem goods and services, and therefore, holistic management that takes into account interdependencies among environmental issues is required.

Keywords Baltic Sea • Invasive alien species (IAS) • Ballast water management • Ecosystem approach to management • Maritime transportation

---

K. Smolarz (✉) • P. Biskup • A. Zgrundo
Department of Marine Ecosystem Functioning, Institute of Oceanography, University of Gdańsk, Al. Marszałka Piłsudskiego 46, 81-378 Gdynia, Poland
e-mail: oceksm@univ.gda.pl; paulina_lemke@wp.pl; oceazg@ug.edu.pl

M. Gilek et al. (eds.), *Environmental Governance of the Baltic Sea*,
MARE Publication Series 10, DOI 10.1007/978-3-319-27006-7_4

4.1 Introduction

Although invasive alien species (IAS) can lead to significant negative environmental impacts in the Baltic Sea, environmental risks of such biological invasions have traditionally not attracted as much attention in the Baltic Sea as in other marine areas. In fact, it is only recently that IAS has received significant attention in formal environmental governance arrangements in the EU and the Baltic Sea region.

The assessment and subsequent management of risks and problems associated with biological invasions are fraught with severe challenges of ecological and sociopolitical nature because of uncertainties with regard to the actual impacts of biological invasions and the rather incoherent architecture of European and regional legislation.[1] Adding to these challenges is also the fact that the biodiversity impacts of IAS can be evaluated as either positive (e.g. introduction of new commercial fish species) or negative (outcompeting of native species) depending on the species and contexts considered. Hence, the high level of uncertainty, complexity and ambiguity linked to IAS risks and impacts, as well as the often significant interrelationships with other environmental issues like eutrophication, overfishing and chemical pollution, may all hinder or delay the development and implementation of relevant legislative acts and management options. It should also be underlined that biological invasions in marine ecosystems often lead to large-scale risks and consequences. This means that the effectiveness of IAS mitigation measures is inherently dependent on cross-border and international cooperation.

This chapter describes and analyses biological invasions as examples of a relatively new and crosscutting domain of environmental governance. The main focus is on identifying and analysing the main barriers and opportunities in the governance of Baltic Sea IAS, as well as on identifying possible ways to achieve environmental governance improvements. The study is based on a 5-year (2009–2014) case study undertaken within the research project 'Environmental Risk Governance of the Baltic Sea' (RISKGOV), where the main emphasis was on alien species (AS) and IAS. The first phase of this study relied on an extensive literature review and database searches, both national and international. The second phase involved discussions and interviews with a number of stakeholders to further explore the issues identified as significant and provide broader input and additional information. Representatives from the following five groups of stakeholders were interviewed (19 interviews in total): officials representing governmental organisations (GOs) (European Commission (EC), HELCOM, national government authorities; five), industry (International Maritime Organisation (IMO), port authorities; three), representatives of non-governmental organisations (NGOs) (global and regional environmental organisations; three), experts from academia (four) along with independent experts and professionals (representatives of various institutions and organisations

[1] Commission staff working document. Executive summary of the impact assessment. European Commission, Brussels, 09.09.2013: http://eur-lex.europa.eu/legal-content/EN/TXT/PDF/?uri=CELEX:52013SC0322&from=EN

advising on environmental issues; four) from Poland, Sweden, Lithuania, Finland and Belgium. The interviews followed an analytical framework covering the broad topics of the IAS framing (definitions used, IAS impact on the environment and humans, etc.), regulatory frameworks, risk assessment and management, processes of communication, scientific uncertainty and disagreement as well as information relating to the role of the respondents' institutions and their responsibilities. A thematic analysis of data was subsequently performed, where emphasis was placed on qualitative meanings as opposed to quantifying data. Furthermore, these findings were supplemented by participatory observations made during relevant conferences, workshops and meetings.

The chapter is organised as follows: Sect. 4.2 describes the concept of alien species and invasiveness and discusses the consequences of introducing such species in the Baltic Sea region. Section 4.3 discusses uncertainty in risk assessment as the main challenge for IAS management. Section 4.4 presents the main legal frameworks related to bioinvasions. Section 4.5 focuses on the framing and implementation of EAM. Section 4.6 presents conclusions and recommendations pertaining to IAS management.

4.2 Framing of the Problem

There are a number of definitions of alien species, invasive species and invasive alien species. If not stated otherwise, within this chapter we concentrate on IAS. The most common definition of IAS is the one proposed by the Convention on Biological Diversity (CBD; COP 6 Decision VI/23[2]), namely, 'alien species whose introduction and spread threatens ecosystems, habitats or species with economic or environmental harm'. The process of IAS invasion is referred to as bioinvasion (or biological invasion). There are several routes to IAS introduction, both natural and man-made, but ballast waters used in maritime transportation are currently seen as the main source of the problem because of the fact that maritime transportation has increased enormously not only worldwide but also in the Baltic Sea region (HELCOM 2009a, b; Olenin et al. 2010). Within the Baltic Sea, despite its relatively small size, ship traffic is one of the densest in the world, both in terms of ship numbers and tonnage. Based on information from the HELCOM AIS (Automatic Identification System) database, more than 50,000 vessels cross the Danish straits per year, and at any time, approximately 2,000 vessels can be found in the Baltic Sea.[3] Therefore, according to the Directorate-General for Maritime Affairs and Fisheries (DG Mare 2012), the Baltic Sea is one of the sea basins with the greatest environmental pressure due to shipping activities.

Globally, IAS is considered to be one of the greatest threats to marine biodiversity, entailing both direct and indirect consequences for ecosystems with various

[2] http://www.cbd.int/decision/cop/?id=7197

[3] http://www.helcom.fi/press_office/news_helcom/en_GB/Ship_traffic_stat/

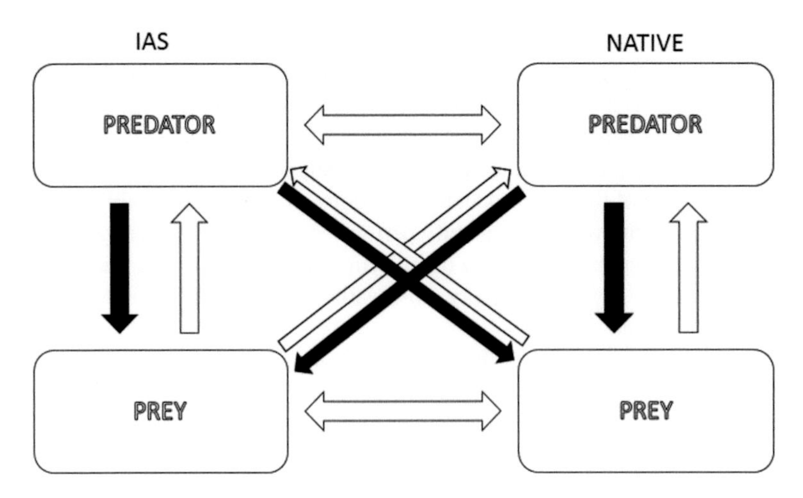

Fig. 4.1 Simplified food-web regulation via three main mechanisms: top-down by population at next trophic level (*dark arrows*), bottom-up by the presence of organisms they feed on (*white arrows*) and horizontal, for example, competition (*double arrows*)

economic and social costs (Born et al. 2005). Zavaleta et al. (2001) describe two basic mechanisms in which food-web interactions can be distorted by IAS introduction, mainly due to the fact that ecosystems usually do not have self-protecting mechanisms against harmful effects of invaders. Identified top-down mechanisms (Fig. 4.1) are linked to the presence of higher level consumers (i.e. grazers or predators) regulating the size of population they feed on, while bottom-up mechanisms regulate the population size of these higher level consumers through the availability of food resources (amount of primary producers or lower level consumers (Zavaleta et al. 2001). The complexity of relations is intensified by the fact that organisms often feed at more than one trophic level. Horizontal mechanisms are mainly linked to competition, for example, for resources, light, physical space and food. However, IAS can affect native species in a number of other ways too, namely, through cross-breeding and the introduction of pathogens. In multiple-invaded ecosystems, therefore, both the horizontal and vertical relationships between different trophic levels become much more complex and uncertain (Bull and Courchamp 2009).

Worldwide IAS-related environmental impacts inflict massive economic costs on fisheries, industry and other human activities (Shine et al. 2010), although the economic evaluation of biological invasions is a difficult task. According to Born et al. (2005), most current studies have methodological shortcomings, mainly due to the fact that they are ex-post impact assessments and insufficiently address uncertainty issues. Furthermore, calculated expenses include mostly direct costs like the damage of harbour infrastructure caused by fouling organisms and usually exclude indirect economic losses such as biodiversity and habitat change, impacts on endemic species and decreasing genetic diversity/identity of local populations. Some introductions, however, can be regarded as beneficial for ecosystems, which adds to the

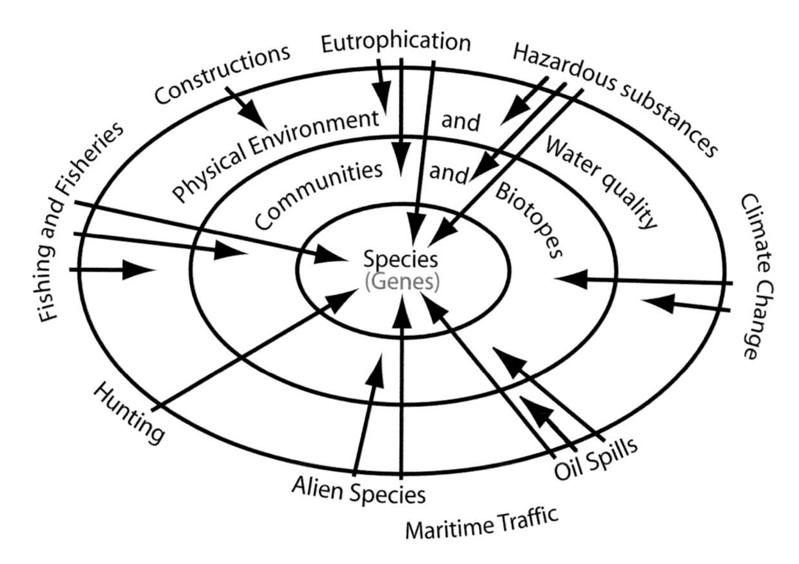

Fig. 4.2 Most important biodiversity threats identified in the Baltic Sea region (Modified from HELCOM 2009c)

complexity of the problem. In such cases, AS (through the processes of primary succession) increase species richness and taxonomic diversity in a given area and increase and strengthen functional diversity and/or provide new services and goods (Bonsdorff 2006; Kotta et al. 2006).

It must be emphasised that the Baltic Sea is 'a sea of aliens' (Leppäkoski et al. 2002), since almost all marine organisms present today invaded the area sometime over the last 10,000 years. However, as it was mentioned previously, the invasion rate has accelerated enormously since the 1950s mainly due to human activities (such as maritime transportation and habitat change). Moreover, synergistic effects of factors such as pollution, eutrophication, overfishing and climate change stimulate new invasions (Fig. 4.2). Changes in environmental conditions and human pressure have on the one hand resulted in the elimination of highly specialised or sensitive native species from the Baltic ecosystem, while on the other hand they have generated and opened niches that have been quickly inhabited by new species. To some extent, the process of bioinvasion can be regarded as positive when species richness and taxonomic diversity are considered. For example, studies on the widespread Baltic invasive polychaete *Marenzelleria* spp. have shown that although this species has become dominant, it has not adversely affected deeper benthic communities, since it fills an 'empty' niche. However, in shallow waters this has not been the case, and *Marenzelleria* spp. competes with the native polychaete worm *Hediste diversicolor* and affects the abundance of the amphipod *Monoporeia affinis* (Didžiulis 2006; Kotta and Olafsson 2003).

Other positive consequences of alien species introduction include human benefits from both AS and IAS which has actually polarised public perceptions about bioinvasion. For example, *Eriocheir sinensis* (Chinese mitten crab), a famous delicacy in Chinese cuisine, invaded the North Sea in the early twentieth century. It quickly became abundant and established self-sustaining populations in German waters disturbing fishery and aquaculture and, among other things, increasing bank erosion. The monetary impact caused by this invader in German waters alone was estimated at approximately 80 million Euros from its advent locally (Gollasch 2011). Recently, however, particularly in the absence of fish, the mitten crab has been harvested in northern Germany and sold in Europe at a price of € 5–8 per kg (Woke 2012). Additionally, it is also seen as a new source of crab for the expanding Chinese market, where the value of sold crabs was estimated as high as € 3–4.5 m between 1994 and 2004 (Gollasch 2011). Thus, what at first seemed to be an adverse alien species turned out to be a new profitable ecosystem good.

Generally, it is estimated that over 120 non-native aquatic species are present in the Baltic Sea, about 80 of which have established viable, self-reproducing populations in at least parts of the region (Battle 2009). However, documented data on environmental impacts is available for only 33, and just 4 species have shown strong impacts on native communities and ecosystem functioning (Battle 2009; Olenin et al. 2010). For example, the invasive dinoflagellate *Prorocentrum minimum* (Pavillard) Schiller was seen to potentially spread and cause significant impacts on plankton communities, habitats and ecosystem functioning (Olenin et al. 2009). One of the most recent examples of bioinvasion in the Baltic Sea was the introduction of a comb jellyfish *Mnemiopsis leidyi* (Kube et al. 2007), one of the world's top 100 worst IAS due to its predatory success, partly attributable to its effective foraging strategy (Colin et al. 2010). Nonetheless, unlike in the Black Sea, where the introduction of this jellyfish led to the collapse of the whole ecosystem (Kideys 2002), this species does not seem to pose any direct threat to the Baltic Sea fishery (Jaspers et al. 2011). In other words, it has to be stressed that no significant degradation of local ecosystems with adverse socioeconomic consequences has taken place in the Baltic Sea as a result of IAS. For that reason, AS and IAS have received far less attention than other stressors affecting biodiversity in the Baltic Sea. In fact, many stakeholders consider the bioinvasion problem in the region as relatively minor. What is, however, dangerous for the Baltic Sea ecosystem is a possible synergistic effect of multiple environmental pressures and IAS. Consequences of such multiple pressures are unknown and may negatively affect ecosystem services and functions.

4.3 Uncertainty in Risk Assessment

The above-mentioned dual nature of AS causes a number of disagreements among stakeholders, as well as uncertainties that are mainly due to the lack of reliable and conclusive scientific data on different aspects of invaders. The absence of information necessary for an adequate risk analysis – required for developing a regional policy instrument of marine governance – is regarded as the main obstacle to effective risk assessment and IAS management in the Baltic Sea (Lemke et al. 2010;

Leppäkoski and Gollasch 2006). Long-term data, made available on a centralised platform, on potentially invasive species regarding ecology, introduction routes and recent changes in ecosystems, are essential for developing multiple risk scenarios and appropriate management options for the Baltic Sea. Moreover, better knowledge about possible consequences of IAS is of great importance, since such knowledge is essential for developing early warning systems, constructing target IAS lists and undertaking cost-benefit analyses of management options and environmental status assessments (Olenin et al. 2010).

Targeted monitoring programmes for nonindigenous species are limited and often restricted to a few invasive species in selected areas. Consequently, relevant data is collected mostly as a result of incidental recordings of IAS during already ongoing biological monitoring. Since the monitoring objectives of such tasks are different than observations meant specifically for IAS introduction, the skills of the people gathering this data and the data quality cannot be guaranteed. Obtained records often appear to be of low quality and should be supplemented by targeted monitoring in high-risk areas of both invasive nonindigenous and indigenous species (EC 2011). Unless better monitoring is in place, there will be no reliable basis for providing advice to management, which would be a major obstacle for ecosystem management of IAS regulation/management in the Baltic Sea. One example of an international project that aimed at improving the prevention of pollution, including IAS spread, from maritime transport was Baltic Master II, a strategic European Union project for the Baltic Sea region. Some improvements can be seen worldwide in recent assessment-management interactions because new programmes have started to deal with uncertainty via targeted monitoring and surveillance, the gathering of new data (e.g. on the biology and ecology of potentially invasive species), the routes and mechanisms of bioinvasion as well as the identification and monitoring of vulnerable sites/routes. The major focus of this development seems to be facilitating the implementation of global 'mitigating' regulations such as the International Convention for the Control and Management of Ships' Ballast Water and Sediments (the Ballast Water Management Convention; BWMC),[4] particularly since the lack of data can be seen as a potential problem for such implementation. Environmental governance, the precautionary principle and EAM are responses to scientific uncertainties and sociopolitical controversies. These approaches are held forward as fruitful alternatives to executive risk-based and sector-restricted regulation, as discussed later in this chapter. However, in the Baltic Sea, IAS is still not regarded as an

[4] The BWM Convention (BWMC 2004) is a voluntary agreement of the International Maritime Organisation (IMO), the latter belonging to the family of UN organisations and shipping industry (in particular, the International Chamber of Shipping, the International Association of Independent Tanker Owners and classification societies), to directly address the main vector of alien species introduction. The convention gives specific guidelines regarding risk assessment and management plans. BMWC will be brought into operation in the near future (HELCOM 2013), namely, when at least 30 countries with 35 % of the registered tonnage of the world merchant fleet have ratified the convention. See: http://www.imo.org/About/Conventions/ListOfConventions/Pages/International-Convention-for-the-Control-and-Management-of-Ships%27-Ballast-Water-and-Sediments-%28BWM%29.aspx

important and urgent issue as in other marine and freshwater basins (e.g. Caspian and Mediterranean Sea, Great Lakes). Only quite recently have IAS received significant attention within EU policy and governance structures (e.g. Shine et al. 2010).

4.4 Bioinvasions as a New Domain of Environmental Governance

Due to the global nature of IAS and the increasing number of vectors of transportation, unilateral action by a single stakeholder (e.g. one nation state), even if aligned with the precautionary principle, is usually insufficient to prevent introductions of AS in marine ecosystems (Shine et al. 2010). Moreover, IAS-related environmental risks do not respond to management measures in the same way as other forms of environmental threats. While, for example, existing threats from chemical pollution and eutrophication can be to some extent decreased if appropriate actions are taken, the risk of biological invasion in marine ecosystems can be reduced only by adopting precautionary measures, since control or eradication of once established species is generally regarded as impossible (EC 2011). Genovesi (2007) has given examples of partially successful cases of control and eradication of IAS from marine waters, but so far none have been recorded in Europe (Genovesi 2005). Hence, preventing the introduction of AS is believed to be the best and most cost-effective option (Olenin et al. 2010).

Horizontal (between different stakeholders and sectors) and vertical (international, regional and local) cooperations are essential for developing effective approaches to address crosscutting issues like IAS. Many national and international management instruments and technical guidelines already deal with this problem, focusing on plant and animal health, biodiversity conservation, aquatic/marine ecosystems and maritime transportation (Lemke et al. 2010). These instruments can be binding (e.g. EU regulations) or more voluntary in nature (e.g. HELCOM recommendations) and provide the baseline from which countries and regional organisations such as the European Union develop policies and legal frameworks designed to decrease IAS threats.

Globally, the problem of IAS is addressed in two conventions, namely, the CBD (1992)[5] and the BWMC (2004). In October 2010, the parties to the CBD agreed the following under the CBD Strategic Plan 2011–2020 regarding IAS: 'By 2020, invasive alien species and pathways are identified and prioritised, priority species are controlled or eradicated, and measures are in place to manage pathways to prevent their introduction and establishment' (EC 2011). Additionally, the Global Invasive Species Programme (GISP) established in 1997 provides support for implementing

[5] CBD (1992) Article 8(h) requires that, as far as possible, each contracting State should *prevent the introduction of, control or eradicate those alien species which threaten ecosystems, habitats or species.*

Article 8(h) of CBD and contributes extensively to knowledge and awareness of invasive species through the development of a range of products and publications, including the 'Global Strategy on Invasive Alien Species' and 'Invasive Alien Species: A Toolkit of Best Prevention and Management Practices'.

The importance of ballast waters as a vector through which IAS is transported has been pointed out by many authors. Ship traffic is recognised as the most important vector introducing new organisms into marine ecosystems, especially so in the Baltic Sea (e.g. Carlton and Geller 1993; Carlton 1996; David and Gollasch 2008; Leppakoski et al. 2002; Pikkarainen 2010; Shine et al. 2010). Ship traffic is mainly addressed in regulations and policies dealing with IAS problems in aquatic environments. Before Regulation (EU) No 1143/2014 on invasive alien species[6] entered into force in January 2015, there had been no consistent and binding legal act or other comprehensive instrument to solve the marine IAS problem in the European Union. However, it should be said that other existing EU legislation and policies do provide some, albeit partial, solutions.[7] Descriptor 2 of Annex I to the EU Marine Strategy Framework Directive (MSFD) among others specifies that by 2020 'non-indigenous species introduced by human activities are at levels that do not adversely alter the ecosystems' (EC 2011). Here, the objective is to keep IAS populations at a level that does not disturb the ecosystem. However, due to the lack of knowledge, it is often impossible to know when exactly a particular species will significantly alter the ecosystem structure and function.

The EU Communication on 'Our life insurance, our natural capital: an EU biodiversity strategy to 2020' (COM(2011)244),[8] tabled by the European Commission in 2011, and then followed up by a resolution on the EU 2020 Biodiversity Strategy adopted by the European Parliament in 2012, set a specific target to address the issue of IAS and proposed the preparation of a dedicated legislative instrument to tackle the problem (EU Biodiversity Strategy to 2020). In September 2013, a new proposal for a regulation on the prevention and management of IAS was issued by the European Commission.[9] This proposal aimed at solving the problem by establishing a framework for action to prevent, minimise and mitigate the negative effects of IAS on biodiversity and ecosystem services. The document underlined the need for coordinated action and prepared a list of species of special concern to the European Union. It also put forward the need for increasing preventive measures

[6] Regulation (EU) No 1143/2014 of the European Parliament and of the Council of 22 October 2014 on the prevention and management of the introduction and spread of invasive alien species. http://eur-lex.europa.eu/legal-content/EN/TXT/PDF/?uri=CELEX:32014R1143&from=EN

[7] Among others: Treaty on the Functioning of the European Union (TFEU – Lisbon Treaty), Sixth Environmental Action Programme (2001–2010), Decision 1600/2002/EC, Council Regulation No 1198/2006 on the European Fisheries Found (OJ L 223, 15.08.2006), Marine Strategy Framework Directive 2008/56/EC (MSFD), Council Regulation No 708/2007 concerning use of alien and locally absent species in aquaculture (OJ L 168 of 28.06.07).

[8] Communication from the commission to the European Parliament, the council, the economic and social committee and the committee of the regions (COM 2011) http://ec.europa.eu/environment/nature/biodiversity/comm2006/pdf/2020/1_EN_ACT_part1_v7%5B1%5D.pdf

[9] http://eur-lex.europa.eu/legal-content/EN/TXT/PDF/?uri=CELEX:52013PC0620&from=EN

and efficiency, as well as lowering the costs of both damage and undertaken actions (COM/2013/0620final).

The Committee on Fisheries (2013) and the Committee on International Trade (2013) of the European Parliament formulated their opinions on the proposal. The former Committee concentrated on matters that were of relevance to the marine environment or aquaculture. A proposed amendment, among others, was with regard to the list of invasive species of importance to the EU. It was felt that such a list should be kept open and regularly revised and that it should take into account that there are invasive species native to some European regions but alien to others. In other words, the required course of action might differ between Member States (MS). Moreover, it was felt that the public should be kept informed and that a scientific advisory group should be established since scientific advice is seen as a 'key to the successful implementation and oversight of the proposed legislation' (Lemke et al. 2010). Opinions were taken into consideration by the Committee on the Environment, Public Health and Food Safety (2013) and included in the amendments laid down in its draft report on the proposed EC legislation mentioned above.

Finally, the Council adopted a new regulation on IAS and published it in the Official Journal of the European Union on 4 November 2014.[10] The new regulation entered into force on 1 January 2015 with its main objectives being to 'prevent, minimise and mitigate the adverse effects of invasive alien species on biodiversity and related ecosystem services, and on human health and safety as well as to reduce their social and economic impact'.[11] The IAS problem was to be addressed in a comprehensive way through three main types of interventions, i.e. prevention, early detection and rapid eradication. It was also stated that IAS that were already widely spread needed to be managed.

Nevertheless, at present there are a few mechanisms to support synchronisation of approaches between neighbouring countries or countries in the Baltic subregion, including Russia. Moreover, so far there is no common EU ballast water policy. It seems that the EU is leaning towards ratifying and implementing the BWM Convention. According to the 2013 Proposal (2013/0307COD), Member States that identified ballast waters as an important pathway would have to include in their action plans measures of the BWMC (Article 11 of the proposal). However, Member States are not obliged to ratify the Convention by the new EU regulation, as it is stated in the document: 'Action should build on the experience gained in the Union and in Member States in managing certain pathways, including measures established through the International Convention for the Control and Management of Ships Ballast Water and Sediments adopted in 2004. Accordingly, the Commission should take all appropriate steps to encourage Member States to ratify that Convention'.[12] Moreover, it was stressed by the Committee on Fisheries in its draft note that although ballast water and hull fouling are the most significant vectors of AS introduction, only five Member States have ratified the BWMC. The Committee

[10] http://ec.europa.eu/environment/nature/invasivealien/index_en.htm

[11] http://eur-lex.europa.eu/legal-content/EN/TXT/PDF/?uri=CELEX:32014R1143&from=EN

[12] http://eur-lex.europa.eu/legal-content/EN/TXT/PDF/?uri=CELEX:32014R1143&from=EN

suggested that the EU Parliament should persuade all the coastal Member States to endorse the Convention, or even that the EU Commission should consider legislative action in this field. Since Member States differ in their perception of IAS, the framework should be developed with the aim of establishing common objectives, terminology and procedures to prevent bioinvasions in the marine environment, as well as controlling measures for sustaining or restoring marine biodiversity.

In the Baltic Sea region, national legislations and policies relating to IAS risk assessment and management are being developed. Risk analysis is one of the most important activities that are necessary to plan appropriate, science-based and cost-effective management options. Risk assessment is *a logical process for assessing the likelihood and consequences of specific events, such as the entry, establishment, or spread of harmful aquatic organisms and pathogens. Risk assessment can be qualitative or quantitative, and can be a valuable decision aid if completed in a systemic and rigorous manner* (MEPC. 162(56)).[13] Marine biosecurity risk assessments follow standardised risk procedures and in case of IAS have been previously based on frameworks and concepts of general ecological risk assessment (Leppäkoski and Gollasch 2006). HELCOM's recommendations and the BWM Convention are the most important frameworks for regulating IAS. Although neither of them are legally enforceable, they constitute a significant platform for dialogue on objectives, strategies and measures among all Baltic countries (Fig. 4.3). The BWMC's and IMO's Marine Environment Protection Committee guidelines are regarded as the most important reference documents regarding key principles defining the nature and performance of IAS risk assessment. However, due to the specific ecological and hydrological characteristics of the Baltic Sea, such as its relatively small size, shallow depth and brackish waters, not all of the management options proposed by IMO could be put into practice in the Baltic Sea (Gollasch and Leppäkoski 2007).

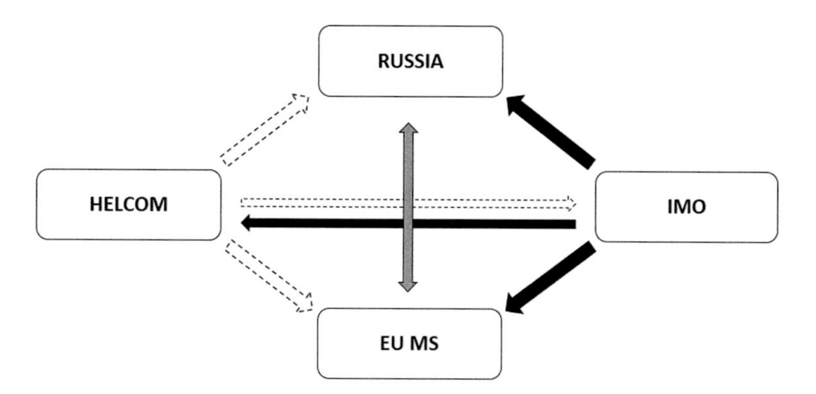

Fig. 4.3 Interactions and ways of communication between main actors involved in IAS decision-making in the Baltic Sea region. *Arrows*: *black*, BWMC; *striped*, recommendations for BWMC ratification and acting according to the Baltic Sea Action Plan (BSAP); *grey*, tension due to different IAS perceptions and priorities occurring within and between riparian states

[13] http://www.imo.org/blast/blastDataHelper.asp?data_id=19689&filename=162%2856%29.pdf

For example, there are no specific procedures to assess risks within – instead of between – biogeographic regions proposed in the IMO Guidelines G7. For that reason, strategies to handle the interregional spread of IAS populations and effective methods to carry out surveys within the Baltic Sea were developed by HELCOM. In October 2013, guidelines for the contracting parties of OSPAR and HELCOM on the granting of exemptions under the BWM Convention, regulation A-4, were adopted by the 2013 HELCOM Ministerial Meeting. This document was jointly developed by the Helsinki and OSPAR conventions, in accordance with Art. 13 (3) of the BWMC, to provide a harmonised procedure for the issue of exemptions from Regulation B-3 (Ballast Water Management for Ships) and Regulation C-1 (Additional Measures) under Regulation A-4 and 'to ensure that exemptions are granted in a constant manner that prevents damage to the environment, human health, property or resources' (HELCOM 2013).

Two recent HELCOM projects have also developed guidelines relating to implementation of IAS regulations and management. The first one aimed at giving the contracting parties the option to test, develop and implement the proposed, harmonised system for granting exemptions.[14] The second one, which was approved by the European Union, studied the harmonisation of the BWM Convention and Marine Strategy Framework Directive monitoring needs on alien species.[15] One of the most relevant outcomes of HELCOM's actions on IAS risks was the development of the Baltic Sea Regional Project (BSRP) which aimed at creating conditions for applying an ecosystem approach to manage the Baltic Sea and sustain its biological productivity. This is regarded as a basis for developing subsequent programmes and management strategies focused on improving the status of the Baltic Sea environment (Thulin 2009).

According to the BWM Convention, all ships have to have ballast water cleaning systems in order to utilise ballast waters. This precondition has to be fulfilled within 2015–2016,[16] by which time the required number of countries is expected to have ratified the Convention. While all countries that have ratified the Convention have to implement it within 12 months, most countries are yet to work on their implementation plans. If a country has ratified the BWM Convention, it needs to ensure that all ports in the country are prepared for BWM provisions. A detailed implementation plan is needed that will address issues such as (1) ballast water- and sediment-mediated bioinvasion risk assessment, (2) ballast water receiving infrastructure in the donor areas and ballast water treatment systems on ships, (3) uncertainties regarding investment needs and costs and (4) definition of responsibilities.

As Lemke et al. (2010) and Kern (2011) argue, the development of law and its implementation strictly depend on the policies of individual states in the Baltic Sea region (BSR). It is believed that EU's centralised structure and the weak political initiatives of some states (particularly new EU members) stand in the way of

---

[14] http://helcom.fi/helcom-at-work/projects/completed-projects/aliens-3

[15] http://helcom.fi/helcom-at-work/projects/balsam/

[16] The status of the BWMC can be checked at: http://www.imo.org/About/Conventions/StatusOfConventions/Pages/Default.aspx

developing a regional IAS approach focused on the entire Baltic Sea region. Moreover, governing the Baltic Sea region depends not only on decisions made by national governments of EU Member States but also on EU-Russia agreements and Russia's bilateral relations with individual EU Member States (Kern 2011). Furthermore, Russia is expected to increase marine transportation from and to St. Petersburg, and the lack of binding agreements obliging Russia to act in accordance with the available recommendations can hamper valuable IAS-related initiatives on a Baltic Sea regional scale.

Apart from the BWM Convention, there are a number of voluntary IAS initiatives on various levels. At the international level, there are the pan-European International Council for the Exploration of the Sea, Code of Practice on the Introductions and Transfers of Marine Organisms and the Council of Europe (Bern Convention). There are also country-based individual voluntary engagements that have pathway codes not covered by existing regulatory frameworks for the Baltic Sea area. Although all the mentioned documents and actions constitute valuable and useful rules, compliance is voluntary and does not impose executive measures. At the same time, although shipping was seen as the main vector of IAS several decades ago, public awareness has been raised only recently. Public debates and campaigns have been aimed at eliciting pro-environmental behaviour and encourage voluntary actions among different stakeholders. Voluntary measures can play multiple roles: raise awareness, create social pressure, develop technological innovations, leverage/disseminate best practices, accelerate regime changes or fill regulatory gaps (Harrison 2001). Bussière and Fratzsher (2008) say that 'for a given degree of risk aversion, there is a unique combination of the forecast horizon and of the probability threshold that maximises the policy maker's preferences, yielding the best possible model from a policy perspective'. The outcome of a political process could therefore be a 'continuous improvement process' policy, which aims at compromising the interests of different stakeholder groups, whereas sustainability goals in terms of meeting economic viability and preserving ecosystem goods and services are often not achieved. Nevertheless, due to constant changes in policy objectives, actual measures taken are a patchwork of attempts to cope with change and complexity.

4.5 Framing and Implementation of the Ecosystem Approach to Management (EAM)

As a consequence of the increasing use of marine resources, concerns about the health of the seas surrounded by large human populations are growing. Thus, far-reaching initiatives to control human activities are being developed. For sustainable management of the marine environment through various management strategies, for example, a maritime security strategy, policy development must address socioeconomic and environmental aspects. However, there is a trade-off between short-term profit and long-term sustainability, and it seems that some of the key stakeholders and management agencies often tend to be interested in the former while others, such as numerous environmental agencies, the latter. Christensen et al. (1996) have listed a number of obstacles contributing to the imbalance between short-term solutions and long-term intergenerational sustainability. These include (1) inadequate available information and scarce knowledge about ecosystem structure and functions, (2) centralisation of management and (3) public perception that social and economic arguments outweigh ecological ones. Thus, ecosystem management and EAM itself were seen to be 'silver bullet' solutions to overcome these obstacles. Yet almost two decades later, despite a focus on EAM, these obstacles still remain. As stated earlier, EAM today is acclaimed worldwide not only in science but also in politics as a means of integrating social, economic and ecological objectives. Consequently, the approach is believed to facilitate sustainable development of marine and coastal areas (e.g. Backer et al. 2010; CBD 2002; Curtin and Prellezo 2010). According to Farmer et al. (2012), one of the most important steps in EAM is the development of conceptual models that capture a wide range of people's perceptions about how the system works. The DPSIR (Drivers, Pressures, State change, Impact and Response) approach has been proposed as a modelling approach aimed at shaping environmental sustainability (Atkins et al. 2011; Elliott 2002, 2003). If one examines IAS from a DPSIR perspective, it becomes clear that Drivers facilitating IAS spread in the Baltic Sea include significant environmental risks such as maritime transportation, eutrophication, overfishing and low biodiversity. The Pressures emanating from increased navigation, ballast water discharge, fouling and aquaculture have the potential to change the status of the biological system (IAS interactions with native species). As a consequence, IAS may have impacts at different levels of the ecosystem (native species, biodiversity change) and affect goods and services of interest to mankind (fish stocks, tourism, water quality, health issues). In turn, such a State change needs a Response at the socioeconomic, technological, administrative and legislative levels. It requires controlling the onset of IAS or at a later stage eradicating an already established IAS, with the latter not being feasible in marine environments. Importantly, a precautionary approach appears to dictate the priorities of a top-down approach to controlling IAS (e.g. through the IMO's ballast water management). However, Shine et al. (2010) argue that the assessment of IAS impacts should generally begin at the local level, in 'hot spots' and 'stepping stone areas'. Those are usually marinas and port basins, or areas of special interest like marine protected areas (MPA). Local assessments can be further integrated into evaluations at the next subregional spatial level (e.g. Gulf of Finland in the Baltic or Adriatic Sea in the Mediterranean) or at the regional sea level (EC 2011).

The EAM approach has already been promoted by the Global Invasive Species Programme (GISP) as an effective management approach against biological invasions (CBD Newsletter, 2009–2007). Such a multi-sector and multi-tiered approach is of particular value in areas where eradication of IAS is not the primary goal. EAM also suggests community involvement in developing management processes and includes recipient ecosystem characteristics. The ecosystem approach as described in CBD Decision V/6 was also proposed as an appropriate framework for the assessment of planned action and policies with regard to biodiversity considerations for isolated ecosystems (Genovesi and Shine 2004). Out of 12 principles adopted for the application of EAM, at least 4 are of particular interest when it comes to bioinvasions in marine ecosystems. Those are decentralisation of transboundary management, preservation of native ecosystem structure and function, consideration of the economy of the region and multi-sector involvement in the management of IAS (Shine et al. 2000). However, EAM requires good multidisciplinary knowledge about the area, something that is often not available for many marine areas. Often the best available knowledge is not sufficient to take fully informed decisions. Yet it is used.

One example of a continuous and interactive step-by-step approach towards ecosystem-based management is the global Marine Spatial Planning (MSP) network that has been introduced using key components of EAM (Ehler and Douvere 2009). Nevertheless, despite many attempts to move MSP beyond the conceptual level, there are still difficulties with regard to combining conservation with human activities so that marine resources are used sustainably.

The EAM concept is present in some IAS-related EU policies. An example is NATURA 2000 where EAM is linked to reinforcement of appropriate assessments for NATURA 2000 sites. Moreover, the ecosystem approach is also the flagship concept in the most influential EU document on the marine environment – the Marine Strategy Framework Directive (MSFD) adopted in 2008 (Bojārs 2012). The MSFD seeks a comprehensive approach that would combine effective environmental protection with sustainable use of marine resources, hence addressing the needs of society. The management framework for marine protection under MSFD is encapsulated in the section on Marine Strategies, which 'shall apply an ecosystem-based approach to the management of human activities' (ibid.). This provides the legal basis that places EAM at the heart of this new obligation placed on EU Member States to address concerns around marine management. However, the Directive does not fully define the sustainability concept nor provide measurable goals specifying future processes and outcomes required for sustainable development. Moreover, it allows for divergent views on what sustainability encompasses and how it can be made operational (Farmer et al. 2012).

The ecosystem approach is also applied in the HELCOM and OSPAR frameworks. In HELCOM BSAP, alien species constitute a factor influencing biodiversity loss. OSPAR has gone even further by developing strategic goals and specific indicators (HELCOM 2010). Another important framework addressing an ecosystem-based approach to resource management is the above-mentioned Baltic Sea Regional Project (BSRP), prepared in collaboration with HELCOM and the Global Environmental Facility (GEF; HELCOM 2006).

Several challenges connected to the implementation of EAM remain such as varying definitions of the approach, lack of cooperation and stakeholder participation, communication concerns and problems with integrating knowledge-based advice across scientific disciplines (Arkema et al. 2006; Barnes and McFadden 2007). In our case study, the major obstacles with EAM implementation pertained to mental obstacles and structural challenges, underfinancing, lack of understanding of ecosystem complexity and dynamics, crosscutting issues and environmental interdependencies and lack of dialogue within and between different stakeholder groups. As a number of authors have argued, too little is being done at the governmental level in terms of justifying scientifically the need for regulatory measures and/or prioritisation of management interventions and, more importantly, in national structures and regional institutions responsible for the implementation of preventive measures (Lemke et al. 2010).

The framing of EAM (with definitions of the concept themselves varying quite substantially) among and within different actor groups dealing with IAS has been polarised. Moreover, the concrete meaning of EAM remains incompletely understood and also implemented in the daily work of many stakeholders (Lemke et al. 2010). Only environmental organisations, scholars and other experts seem to be familiar with the meaning of the concept, but they tend to regard EAM as another way of describing their daily work. The EAM model was also seen to be the only holistic way to influence ongoing crosscutting environmental issues. However, this requires breaking down mental and practical barriers that are present in many sectors of the economy and administration (ibid.).

For many actors, the concept serves as a bridge between science, policy and management and is therefore a useful tool in planning, developing and managing activities addressing societal needs and protecting the marine environment. For some stakeholders, it is also regarded as a 'silver bullet' to solve all Baltic Sea problems, which was perceived of as unrealistic by many environmental agencies. It is also important to note that many interviewed representatives from the shipping industry were unfamiliar with EAM and not convinced after being briefed on the concept (Lemke et al. 2010). These stakeholders tend to believe that EAM represents a multi-scale and multidisciplinary challenge, if not an unachievable utopia, and many stated it was irrelevant to what they actually do. This scepticism might be due to the fact that the concept includes adaptive management, acknowledges gaps in ecosystem knowledge and uncertainties, considers multiple external influences and strives to balance diverse societal objectives, but consequently lacks clear goals and tools to implement it in practice.

Many stakeholders feel that EAM lacks an integrated strategy that considers the heterogeneous characteristics of marine space and resolves conflicts among consumers and the natural environment. The challenge of EAM can, however, be turned into an opportunity to improve the foundation of EAM. For this to happen, risk assessment and risk management will have to undergo significant changes, since a variety of institutions need to adapt to new ways of giving, using and implementing recommendations in their daily operating and decision-making processes in accordance with EAM.

4.6 Future European IAS Policy: Conclusions and Recommendations

Increased awareness and legislative mandates pertaining to sustainable development need to be incorporated in the management approaches of natural resource management agencies. Most approaches that seek to address sustainable development have been criticised for focusing on short-term gain rather than long-term environmental, economic and social profits. As mentioned above, far-reaching environmental pressures, associated with the degradation of resources and loss of ecosystem services in the Baltic Sea due to eutrophication, overfishing, pollution, climate change and IAS, together with increasing maritime transportation, are expected to accelerate the spread of AS in the near future. A number of these environmental pressures will further modify ecological processes in ways that boost the societal and environmental impacts of invasive species (Pyke et al. 2008). IAS is a multifaceted problem characterised by complexities in risk assessment and management strategies. Despite structural complexity, functional connectedness and dynamism of an ecosystem, management options often treat different risks as independent issues, while pressures are addressed in separate policies. In the case of IAS, only combined cross-sectoral action could overcome difficulties related to uncertainties in risk assessment and management. We believe that future marine policies covering IAS should include horizontal and vertical interactions between different environmental pressures and their consequences. Currently, only the IMO BWM Convention covers IAS spread via ballast water exchange. Moreover, this policy responds only to ship-mediated species introduction. Furthermore, it is not ratified yet, is not regarded as robust enough to cover different environmental risks and is not detailed enough to be appropriate for Baltic Sea conditions (as explained in Sect. 4.4). There is a need for regulatory measures addressing both various aspects of invasiveness and associated environmental risks, as well as the specificity of the Baltic Sea ecosystem and the political situation that surrounds it.

We believe that IAS management in the Baltic Sea region is ineffective mostly due to a heterogeneous legislative system. First of all, Baltic Sea countries have international obligations to address IAS, principally according to the CBD and BWM Conventions, which are of a general character. Major concerns for future IAS management within the Baltic Sea region are (1) the plethora of EU Directives (i.e. EU Habitats & Species Directive, Water Framework and Marine Strategy Framework Directives, Directive 2014/89/EU on Maritime Spatial Planning) each implemented to a lesser or greater degree depending on Member State's policies and (2) the number of statutory and other bodies concerned with implementing EU directives and agreements. Thus, the major challenge for the coming years will be promoting Integrated Coastal Zone Management, linked to and based on existing legal frameworks. An integrated governance approach in the Baltic Sea region is a difficult task given the presence of new European Union members in its southern and eastern parts which are also greatly influenced by the Russian Federation. Certain differences in, for example, the understanding of democracy, awareness of environmental issues and approach to citizens' engagement in public debates and NGO activities are observed among these riparian states. Moreover, the economic situation differs widely between the countries in the Baltic Sea region and consequently the impact of their industrial and agricultural sectors on the marine environment. Furthermore, transport policies are caught between demands for improved mobility on the one hand and the concern for environmental protection on the other.

The EU IAS policy must be seen in the context of commitments at global, EU and regional levels. It was suggested that a dedicated IAS directive was the most effective way to provide a flexible framework with minimum standards for IAS action in the EU (Shine et al. 2009). Finally, a new dedicated legislative instrument for IAS management was adopted by the Council and entered into force at the beginning of 2015.[17]

Different Member States have been developing different approaches. The lack of a coherent and coordinated approach has hampered the effectiveness of these initiatives, and hence, IAS-related problems continue to increase. Currently, there is a lot of preparation being done at EU and HELCOM levels to help effectively implement the BWM convention. Some EU regulations together with HELCOM guidelines could be used as examples of coordination and consideration of regional requirements for the implementation of BWMC. EU Member States are obliged to cooperate regionally according to Articles 5 and 6 of the MSFD. Furthermore, they must ensure that marine strategies are coherent and coordinated across marine regions or subregions. A new strategy will be based on planned cooperation at EU and MS levels, and for that purpose, Member States must use, where practical and appropriate, existing regional cooperation structures. They are therefore obliged to follow a common approach to initial assessment, determination of good environmental status, management targets, indicators, monitoring and measures (EC 2011). Regarding shipping, ratification of the BWM Convention is a prerequisite, since it is difficult for the EU to take decisions outside of the IMO framework. One of the advantages of the IMO framework is that it is coherent. It also provides for the control of vessels not registered in the EU, but transiting EU waters (Farmer et al. 2012). The EU's objective is to provide a flexible framework with minimum standards based on the precautionary principle and IAS policy so as to ensure coherence with upcoming instruments and emerging solutions (Shine et al. 2010). Preventive measures are also necessary, as well as taking into account interactions with other environmental risks such as eutrophication and overfishing occurring in the Baltic Sea. Preventive measures are diverse and range from education (e.g. public environmental campaigns, workshops, staff training in the maritime industry) to ballast water exchange regulations. Apart from being addressed within maritime transportation and the BWM Convention, the IAS issue should also be considered in other policies dealing with eutrophication, climate change and overfishing since these threats can undermine the conditions under which current IAS policies are developed (especially those where risk assessment is based on species range and invasion pathways). Before preparing any management strategies, effective policies dealing with

[17] http://ec.europa.eu/environment/nature/invasivealien/index_en.htm

complexity should consider (1) interrelated and crosscutting environmental issues, (2) dynamics of marine ecosystems and (3) clear goals and incentives to vigorously put into practice a new legislation (Shine et al. 2010).

Adaptive management is a way to improve the management process for IAS. Adaptive management is one of the concepts central to EAM that is also integrated in MSFD. Adaptive management recognises that long-term management decisions based upon conceptual modelling or knowledge of only a limited part of the system are unwise due to scientific uncertainty about related natural systems (Farmer et al. 2012). Thus, scientific knowledge and understanding of the system are necessary to reduce management uncertainties. Similarly, effective cooperation of all key stakeholders in a marine region is crucial to facilitate desired outcomes of adaptive management plans. Long-term goals should be revised periodically as new scientific information becomes available and as social needs change (ibid.). Developing a new European legislation on IAS not only generates problems and challenges but also provides an excellent opportunity to bring together crosscutting inquiries and reconsider interactions between different environmental threats. HELCOM was therefore proposed to be the main driving force in the implementation of EAM in the Baltic Sea marine area, as well as the coordinating platform for the Baltic Sea regional implementation of the Marine Directive in the Baltic Sea (EC 2011).

In summary, we believe that despite increased activity in some IAS sectors, there is room for more action at the governmental level as has been the case, for example, in Australia and New Zealand. As importantly, more action is needed within nation-states and regional institutions, especially those responsible for the implementation of all the proposed preventive measures. Despite a number of actions taken, the impact has been limited and the spread of IAS continues to take place. To move from conceptualising solutions to translating them into practice requires (1) defining proper goals and objectives concentrating on sustainability, (2) intergenerational sustainability through the harmonisation of ecosystem processes and functions with management actions and (3) switching to adaptive management that by definition requires cooperation between different stakeholders. Moreover, as IAS risks are interrelated with other environmental threats (e.g. eutrophication, overfishing and climate change), invasive species considerations should be integrated into other policies. These policies should include information on the relevant linkages and recognise the interactions and synergistic effects. The implementation of the BWM Convention dealing with ship-mediated species introduction is an important step forward, but it needs to enter into force. Second, the Convention is not robust enough to cover different environmental risks. Finally, practically speaking, the Convention is still not detailed enough to be appropriate for the Baltic Sea given its peculiar conditions. For these reasons, an implementation strategy for each country that has ratified the BWM Convention is necessary. These strategies would have to address issues such as (1) uncertainties linked to specificity of the Baltic Sea environment and ballast water removal and (2) amending the distribution of responsibilities between ship owner and harbours (e.g. financial responsibilities, planning and building infrastructure for receiving ballast water). Additionally, there is a need to introduce new or reinforce existing mechanisms that support a more ecosystem-based governance system. These should ensure not only systematic collection of basic data and the use of the best scientific advice based on robust data but also the introduction of environmental impact assessment as a precondition for new activities. Policy changes and the involvement of stakeholders are crucial to future policy frameworks in order to prevent unwanted introductions and to cover additional vectors of transportations of marine IAS. Voluntary measures are recommended since they supplement frameworks and regulations. Such measures are especially important because high impact policy ideas often lose effectiveness during the political process. We believe that sharing an understanding of EAM with key stakeholders is an important step towards improved environmental governance. We also argue that, for the Baltic Sea, EAM is the only way forward and a good tool for managing risks such as bioinvasions. EAM, the IMO BWMC and international or regional networks could be a solid foundation for developing an integrated system for marine IAS management in the Baltic Sea region.

Acknowledgements This research was funded by the European Community's Seventh Framework Programme (2007–2013) under grant agreement n° 217246 made with the joint Baltic Sea research and development programme BONUS, as well as by the Polish Ministry of Science and Higher Education. Thanks to two peer reviewers for valuable comments on an earlier version of the chapter.

References

Arkema KK, Abramson SC, Dewsbury BM (2006) Marine ecosystem-based management: from characterization to implementation. Front Ecol Environ 4:525–532

Atkins JP, Burdon D, Elliott M, Gregory AJ (2011) Management of the marine environment: integrating ecosystem services and societal benefits with the DPSIR framework in a system approach. Mar Pollut Bull 62:215–226

Backer H, Leppänen JM, Brusendorff AC, Forsius K, Stankiewicz M, Mehtonen J, Pyhälä M, Laamanen M, Paulomäki H, Vlasov N, Haaranen T (2010) HELCOM Baltic Sea Action Plan – a regional programme of measures for the marine environment based on the ecosystem approach. Mar Pollut Bull 60:642–649

Barnes C, McFadden KW (2007) Marine ecosystem approaches to management: challenges and lessons in the United States. Mar Policy 32:387–392

Battle J (ed) (2009) Silent invasion: the spread of marine invasive species via ships' ballast water. WWF International, Gland

Bojārs E (2012) Towards a resilient ecosystem of the Baltic Sea. Legal frame for monitoring, assessment and reporting on the status of marine biodiversity. The brochure produced in the frame of the LIFE+ Nature & Biodiversity project "Innovative approaches for marine biodiversity monitoring and assessment of conservation status of nature values in the Baltic Sea" (Project acronym -MARMONI)

Bonsdorff E (2006) Zoobenthic diversity-gradients in the Baltic Sea: continuous post-glacial succession in a stressed ecosystem. J Exp Mar Biol Ecol 330:383–391

Born W, Rauschmayer F, Brauer I (2005) Economic evaluation of biological invasions – a survey. Ecol Econ 55:321–336

Bull LS, Courchamp F (2009) Management of interacting invasives: ecosystem approaches. In: Clout MN, Williams PA (eds) Invasive species management: a handbook of principles and techniques. Oxford University Press, Oxford, pp 232–247

Bussière M, Fratzscher M (2008) Low probability, high impact: policy making and extreme events. J Policy Model 30:111–121

Carlton J (1996) Pattern, process, and prediction in marine invasion ecology. Biol Conserv 78:97–106

Carlton J, Geller J (1993) Ecological roulette: the global transport of nonindigenous marine organisms. Science 26:78–82

CBD (Convention on Biological Diversity) (2002) Guiding principles for the prevention, introduction and mitigation of impacts of alien invasive species that threaten ecosystems, habitats or species annexed to Decision VI/23. The Hague, April 2002

Christensen M, Norman L, Bartuska AM, Brown JH, Carpenter S, D'Antonio C, Francis R, Franklin JF, MacMahon JA, Noss RF, Parsons DJ, Peterson CH, Turner MG, Woodmansee RG (1996) The report of the Ecological Society of America Committee on the scientific basis for ecosystem management. Ecol Appl 6:665–691

Colin S, Costello J, Hansson L, Titelman J, Dabiri J (2010) Stealth predation and the predatory success of the invasive ctenophore *Mnemiopsis leidyi*. Proc Natl Acad Sci U S A 107:17223–17227

Committee on Fisheries (2013) Draft opinion of the Committee on Fisheries for the Committee on the Environment, Public Health and Food Safety on the proposal for a regulation of the European Parliament and of the Council on the prevention and management of the introduction and spread of invasive alien species (COM (2013)0620-C7-0264/2013- 2013/307(COD)

Committee on International Trade (2013) Draft opinion of the Committee on Committee on International Trade for the Committee on the Environment, Public Health and Food Safety on the proposal for a regulation of the European Parliament and of the Council on the prevention and management of the introduction and spread of invasive alien species (COM(2013)0620-C7-0264/2013- 2013/307(COD)

Committee on the Environment, Public Health and Food Safety (2013) Draft report on the proposal for a regulation of the European Parliament and of the Council on the prevention and management of the introduction and spread of invasive alien species (COM(2013)0620-C7-0264/2013- 2013/307(COD)

Convention on the protection of the marine environment of the Baltic Sea area (Helsinki Convention), Ospar Convention for the protection of the marine environment of the north-east Atlantic, First joint ministerial meeting of the Helsinki and Ospar commissions (jmm), Bremen: 25–26 June 2003: Statement on the ecosystem approach to the management of human activities "Towards an ecosystem approach to the management of human activities". Accessed 14 June 2012

Curtin R, Prellezo R (2010) Understanding marine ecosystem based management: a literature review. Mar Policy 34:821–830

David M, Gollasch S (2008) EU shipping in the dawn of managing the ballast water issue. Mar Pollut Bull 56:1966–1972

Didžiulis V (2006) NOBANIS – Invasive alien species fact sheet – *Marenzelleria neglecta*. Online Database of the North European and Baltic Network on Invasive Alien Species – NOBANIS. Available from: www.nobanis.org. Accessed 20 Aug 2012

EC (European Commission) (2011) Working paper. Relationship between the initial assessment of marine waters and the criteria for good environmental status
Ehler Ch, Douvere F (2009) Marine spatial planning: a step-by-step approach toward ecosystem-based management. Intergovernmental Oceanographic Commission and Man and the Biosphere Programme. IOC Manual and Guides No. 53, ICAM Dossier No. 6. UNESCO, Paris
Elliot M (2002) The role of the DPSIR approach and conceptual models in marine environmental management: an example for offshore wind power. Mar Pollut Bull 44:iii–vii
Elliot M (2003) Biological pollutants and biological pollution – an increasing cause for concern. Mar Pollut Bull 46:275–280
EU Biodiversity Strategy to 2020 – towards implementation. Available from: http://ec.europa.eu/environment/nature/biodiversity/comm2006/2020.htm. Accessed 23 Aug 2012
Farmer A, Mee L, Langmead O, Cooper P, Kannen A, Kershaw P, Cherrier V (2012) The ecosystem approach in marine management. EU FP7 KNOWSEAS Project
Genovesi P (2005) Eradications of invasive alien species in Europe: a review. Biol Invasions 7:127–133
Genovesi P (2007) Limits and potentialities of eradication as a tool for addressing biological invasions. In: Caldwell MM, Heldmaier G, Jackson RB, Lange OL Mooney HA, Schulze ED, Sommer U (eds) Biological invasions, Ecological studies vol 193, pp 385–403
Genovesi P Shine C (2004) European strategy on invasive alien species. Convention on the conservation of European Wildlife and habitats (Bern Convention) Nature and environment, no. 137
Gollasch S (2011) NOBANIS – invasive alien species fact sheet – *Eriocheir sinensis*. Available from: Online Database of the European Network on Invasive Alien Species www.nobanis.org. Accessed 22 Jan 2013
Gollasch S, Leppäkoski E (2007) Risk assessment and management scenarios for ballast water mediated species introductions into the Baltic Sea. Aquat Invasions 4:313–340
Harrison K (2001) Voluntarism and environmental governance. In: Parson E (ed) Governing the environment: persistent challenges, uncertain innovations. University of Toronto Press, Toronto
HELCOM (2006) Report of the second HELCOM/BSRP thematic workshop on the risk related to the ballast water mediated introductions of non-indigenous species in the Baltic Sea
HELCOM (2009a) Overview of the shipping traffic in the Baltic Sea. Available from: www.helcom.fi/stc/files/shipping/Overview%20of%20ships%20traffic_updateApril2009.pdf. Accessed May 2010
HELCOM (2009b) HELCOM activities related to the transfer of alien species in the Baltic Sea. Available from: www.unep.org/regionalseas/globalmeetings/11/inf.08-alien-species-HELCOM.pdf. Accessed Sept 2010
HELCOM (2009c) Biodiversity in the Baltic Sea – An integrated thematic assessment on biodiversity and nature conservation in the Baltic Sea. Balt. Sea Environ. Proc. No. 116B. Available from: http://helcom.fi/Lists/Publications/BSEP116B.pdf#search=ship%20traffic%20stat. Accessed June 2014
HELCOM (2010) Road map towards harmonized implementation. Available from: www.helcom.fi/BSAP/ActionPlan/otherDocs/en_GB/roadmap/. Accessed May 2010
HELCOM (2013) Joint HELCOM/Ospar guidelines on the granting of exemptions under the International Convention for the Control and Management of Ships' Ballast Water and Sediments, Regulation A-4
Jaspers C, Titelman J, Nansson LJ, Haraldsson M, Ditlefsen CR (2011) The invasive ctenophore *Mnemiopsis leidyi* poses no threat to Baltic cod eggs and larvae. Limnol Oceanogr 56:431–439
Kern K (2011) Governance for sustainable development in the Baltic Sea region. J Balt Stud 421:21–35
Kideys AE (2002) Fall and rise of the Black Sea ecosystem. Science 297:1482–1484

Kotta J, Olafsson E (2003) Competition for food between the introduced polychaete *Marenzelleria viridis* (Verrill) and the native amphipod *Monoporeia affinis* Lindström in the Baltic Sea. J Sea Res 50:27–35

Kotta J, Kotta I, Simm M, Lankov A, Lauringson V, Põllumäe A, Ojaveer H (2006) Ecological consequences of biological invasions: three invertebrate case studies in the north-eastern Baltic Sea. Helgol Mar Res 60:106–112

Kube S, Lutz P, Honnef C, Augustin CB (2007) *Mnemiopsis leidyi* in the Baltic Sea – distribution and overwintering between autumn 2006 and spring 2007. Aquat Invasions 2:137–145

Lemke P, Smolarz K, Zgrundo A, Wolowicz M (2010) Biodiversity with regard to alien species in the Baltic Sea region. RISKGOV report to BONUS EEIG Programme; University of Gdansk, Institute of Oceanography, Gdynia. Available from: http://www.sh.se/riskgov

Leppäkoski E, Gollasch S (2006) Risk assessment of ballast water mediated species introductions – A Baltic Sea approach. HELCOM report. Available from: www.helcom.fi/stc/files/shipping/Ballast%20Water%20Mediated%20Species%20Introductions.pdf. Accessed June 2010

Leppäkoski E, Gollasch S, Olenin S (eds) (2002) Invasive aquatic species of Europe – distribution, impact and management. Kluwer Academic Publishers, Dordrecht

Mare DG (2012). Blue growth. Scenarios and drivers for sustainable growth from the oceans, seas and coasts. Third Interim Report

Olenin I, Hajdu S, Wasmund N, Jurgensome I, Gromisz S, Kownacka J, Toming K, Olenin S (2009) Impacts of invasive phytoplankton species on the Baltic Sea ecosystem in 1980–2008. HELCOM Baltic Sea Environmental Fact Sheet, Available from http://helcom.fi/baltic-sea-trends/environment-fact-sheets. Accessed June 2009

Olenin S, Alemany F, Cardoso AC, Gollasch S, Goulletquer P, Lehtiniemi M, McCollin T, Minchin D, Miossec L, Occhipinti Ambrogi A, Ojaveer H, Rose JK, Stankiewicz M, Wallentinus I, Aleksandrov B (2010) Marine Strategy Framework Directive, Task Group 2 report, non-indigenous species. Office for Official Publications of the European Communities, Luxembourg

Pikkarainen A (2010) Maritime traffic and alien species introductions in the Baltic Sea. Available from: www.merikotka.fi/tiedotteet/Pikkarainen_2010_Alien_species_final2. Accessed Sept 2010

Pyke CR, Thomas R, Porter RD, Hellmann JJ, Dukes JS, Lodge DM, Chavarria G (2008) Current practices and future opportunities for policy on climate change and invasive species. Conserv Biol 22(3):585–592

Shine C, Williams N, Gündling L (2000) A Guide to designing legal and institutional frameworks on alien invasive species. IUCN, Gland

Shine C, Kettunen M, ten Brink P, Genovesi P, Gollasch S (2009) Technical support to EU strategy on invasive species (IAS) – recommendations on policy options to control the negative impacts of IAS on biodiversity in Europe and the EU. Final report for the European Commission. Institute for European Environmental Policy (IEEP), Brussels

Shine C, Kettunen M, Genovesi P, Essl F, Gollasch S, Rabitsch W, Scalera R, Starfinger U, ten Brink P (2010) Assessment to support continued development of the EU Strategy to combat invasive alien species. Final report for the European Commission. Institute for European Environmental Policy (IEEP), Brussels

Thulin J (2009) The recovery and sustainability of the Baltic Sea large marine ecosystem. In: Sherman K, Aquarone MC, Adams S (eds) Sustaining the world's large marine ecosystems. IUCN, Gland, pp 63–75

Woke L (2012) Europe's unwanted guest savoured in China. China Daily 2012-10-15. Available from: http://www.chinadaily.com.cn/cndy/2012-10/15/content_15816422.htm. Accessed 22 Jan 2013

Zavaleta ES, Hobbs RJ, Mooney HA (2001) Viewing invasive species removal in a whole-ecosystem context. Trends Ecol Evol 16:454–459

Chapter 5
Governance of Chemicals in the Baltic Sea Region: A Study of Three Generations of Hazardous Substances

Mikael Karlsson and Michael Gilek

Abstract This study zooms in on public governance in the Baltic Sea region of three generations of notorious hazardous substances, namely, PCBs, PBDEs and PFOS/PFOA. Following regulation, PCB concentrations in the Baltic Sea have decreased substantially although they are still above pre-industrial levels. PBDE levels have also decreased in some places, but they too are well above targeted levels, whereas the situation for PFOS and in particular for PFOA has hardly improved at all. In the case of PCBs, while comprehensive measures took long to implement, initial preventive measures were taken early based on the precautionary principle. This contrasts with the cases of PBDEs, PFOS and PFOA, where the burden of proof on policy-makers has been high and hence caused severe delays in policy-making. There has, however, generally been a positive interplay in all three cases between the EU, which has legislated, and HELCOM, which has taken the role of concept and agenda setting. While environment-oriented policies, such as the Ecosystem Approach to Management under MSFD and BSAP, have grown in importance over time, polluter-oriented chemical legislation has been more important when it comes to final decision-making. Nevertheless, the general response has been reactive rather than proactive, and there is no indication that society responds faster today than in the past, at least not given the fact that awareness, experience and knowledge are greater today than a few decades back. Based on that insight, the article discusses various options for improving governance.

Keywords PCB • Brominated flame retardants • Fluorinated substances • Ecosystem approach to management • Precaution

M. Karlsson (✉) • M. Gilek
School of Natural Sciences, Technology and Environmental Studies, Södertörn University, 14189 Huddinge, Sweden
e-mail: mikael.karlsson@2050.se; michael.gilek@sh.se

M. Gilek et al. (eds.), *Environmental Governance of the Baltic Sea*, MARE Publication Series 10, DOI 10.1007/978-3-319-27006-7_5

5.1 Introduction

It is well known that tens of thousands of man-made chemical substances are used by industries in the making of various products in the Baltic Sea area. Hundreds of these, known to have inherent hazardous properties, are emitted and present in the marine environment, both in the sea and in living organisms in and around the sea, including in humans (HELCOM 2010a; Lyons 1999). The negative consequences of these substances on various species have been well documented in the case of the Baltic Sea, including reproductive disorders in marine mammals (Bergman 2007) and imposex in snails (Santillo et al. 2001). In other cases, the long-term consequences are more difficult to interpret, for example, the fact that newborns of mothers on the East Coast of Sweden, who consume a relatively high amount of contaminated fish, weigh significantly less than newborns on the West Coast (Rylander et al. 2000). Most likely, the full picture of consequences of hazardous substances in the marine environment is still emerging.

In fact, when it comes to understanding the total risks of real life exposure to the very complex mixture of hundreds or more industrial chemicals in the Baltic Sea, there are huge information gaps. This is basically due to lack of knowledge and data on properties and exposure conditions for the vast majority of substances (Allanou et al. 1999; Gilbert 2011; Rudén and Hansson 2010) and in particular the adverse consequences of combinations of these substances (Kortenkamp et al. 2009). On top of this complex pollution situation with extreme levels of scientific uncertainty, a complex and fragmented governance system consisting of multi-level, multi-sector and multi-actor interactions escalates the challenges associated with environmental policy objectives. That the Baltic Sea ecosystem in addition is more vulnerable to pollution than most other sea areas (Magnusson and Norén 2012) is not making the task easier.

To cope with the problems and risks of chemical pollution, a number of governance structures and strategies have been put in place, aimed at what has been termed a "Baltic Sea with life undisturbed by hazardous substances" (HELCOM 2007). However, in spite of quite successful mitigation efforts in relation to some pollutants, overall goals are far from being realised (HELCOM 2010a; MMB 2012) and the resulting costs of chemical contamination can be very high (KEMI 2013a; NCM 2004; UNEP 2013). At the EU level, for example, it has been roughly estimated in one study that exposure to endocrine disrupting chemicals costs 13–31 billion Euros annually (Jensen 2014), whereas the costs today of impaired male reproduction have been calculated in another study to reach above 1.2 billion Euros, with variations up and down depending on the assumptions made (NCM 2014). Over time, the political landscape, mitigating measures and environmental governance approaches in place have all evolved. In the 1960s, "polluter-oriented" approaches emerged, commonly focusing on national command and control of point sources, which gradually were complemented with "environment-oriented" approaches, based on broader and ecologically more holistic perspectives (Karlsson et al. 2011). Under the former approach, in the "sphere" of, for instance, chemicals policy, preventive measures were often balanced by compromises based on technological and economic parameters, such as in requirements for best available technology. Such measures still constitute important parts of environmental law in the Baltic Sea region, for example, in legislation on the use of chemical substances, such as the REACH regulation (EC 2006a) of the European Union (EU). In the latter environment-oriented "sphere", the starting point is rather health and ecosystem parameters, which is evident, for instance, in the setting of environmental quality standards and in criteria and indicators for Good Environmental Status under the EU Marine Strategy Framework Directive (EC 2008a).

This development in law and practice has gone hand in hand with new theories on environmental policy, underlining the need for broad governance studies and strategies. While the governance concept has been described in various ways (Adger and Jordan 2009; Kooiman 2003; Pierre and Peters 2005; Young 1994), a common core recognises the transfer of government authority upwards to international institutions, sideways to non-governmental actors and downwards to local actors (Kern and Löffelsend 2008). In the multi-country, multi-sector, multi-risk and multi-stakeholder environment of the Baltic Sea area, broad governance measures are not only taking place but are also considered normatively desirable (Joas et al. 2008), not least when it comes to the prevalent call for an 'ecosystem approach to management' (EAM) (Backer et al. 2010; Curtin and Prellezo 2010; Karlsson et al. 2011; Murawski 2007; Österblom et al. 2010). EAM is integral to the Convention on Biodiversity (UN 1992) as well as in the work of HELCOM, the executive body under the Helsinki Convention (1974), where it has been defined as (HELCOM and OSPAR 2003):

> "the comprehensive integrated management of human activities based on the best available scientific knowledge about the ecosystem and its dynamics, in order to identify and take action on influences which are critical to the health of marine ecosystems, thereby achieving sustainable use of ecosystem goods and services and maintenance of ecosystem integrity". The application of the precautionary principle is equally a central part of the ecosystem approach.

Under this environment-oriented governance approach, which recognises complexities in both natural and social systems, all relevant and interlinked systems and parameters are supposed to be considered across scales and sectors over time and in the light of the precautionary principle. Nowadays, EAM is expressed in both policy, for example, the Baltic Sea Action Plan (HELCOM 2007), and law, for example, the Marine Strategy Framework Directive (EC 2008a).

This case study on public governance of chemicals in the Baltic Sea investigates three key examples from three generations of halogenated organic industrial substances, namely, PCBs (chlorinated), PBDEs (brominated) and PFOS/PFOA (fluorinated). This choice of sub-cases has been made in order to allow for an in-depth analysis of how public governance has evolved under different periods and political systems, with varying degrees of uncertainty and controversies. In doing so, the article describes the co-evolution of risk and governance in each sub-case, with a focus on two key bodies for environmental policy in the Baltic Sea region – the EU and HELCOM. The ultimate aim is to elaborate on potential strategies for improving the fulfilment of environmental objectives at hand. More specifically, the study addresses the following questions:

1. Which measures – such as legislation, recommendations and policies – have the EU and HELCOM taken to manage PCBs, PBDEs and PFOS/PFOA; how do these relate to risk assessment, scientific uncertainty and controversies; and what governance outcomes can be identified?
2. Which governance approaches – such as risk-based polluter-oriented command and control, or EAM, including the precautionary principle – have been applied in each of the sub-cases?
3. What can be learnt from the past and present, and which strategies can be identified for potentially improving risk governance of chemicals in the Baltic Sea region (BSR)?

The focus on EU and HELCOM, based on the insight that these institutions are the most important ones for the governance of chemicals in BSR (Karlsson et al. 2011), means that measures at national or other levels, as well as voluntary measures, will not be studied in any detail. Furthermore, the study centres on industrial chemicals in general and not on specific groups of chemicals or products, such as pharmaceuticals or toys, since the associated risks and regulations in such cases often deserve specific attention. In addition, the emphasis is placed on initial science-policy relations, and on broader policies and general legislation, rather than on the often very detailed and diversified regulations that develop once a problem has been commonly recognised and measures have been institutionalised in society along products' and substances' life cycles.

The study is primarily based on a review of documents and literature, but also on a series of in-depth interviews with a number of stakeholders. The documents studied were peer-reviewed scientific publications on problems and management in the area, so-called grey literature, such as reports and other types of publications from non-governmental and governmental agencies and institutions working on environmental governance in BSR, and political documents. The latter consisted of laws, policies, plans, assessments and various types of documents pertaining to strategies and management tools, mostly at EU and international levels. Examples include:

- EU: the Water Framework Directive (EC 2000), the Marine Strategy Framework Directive (EC 2008a) and the REACH regulation (EC 2006a), together with several technical risk assessment documents (e.g. EEC 1993; European Commission 1994).
- HELCOM: the Baltic Sea Action Plan (BSAP) (HELCOM 2007) under the Helsinki Convention (1974, 1992), several politically adopted recommendations and technical documents and reports for monitoring and assessment.

The in-depth interviews were carried out as part of the broad RISKGOV project and have been documented and reported elsewhere (Karlsson et al. 2011; Udovyk et al. 2010). A total of 22 semi-structured interviews, with open-ended questions, were conducted with scientists, politicians and journalists, as well as other actors in HELCOM institutions and national and EU authorities. For this study, the general results from the interviews are foremost used to structure the analysis and find the broader patterns over time in relation to the sub-cases studied.

In what follows, we present an overview of the development of governance in BSR over time, after which the three generations of sub-cases are detailed. Each result section starts with a problem presentation and a description of the more specific governance measures taken in the EU and by HELCOM, respectively, and is followed by an analysis of trends and approaches. In doing so, the policies, laws and other tools discussed are brought up chronologically and related to each case. The article ends with a discussion on potential roads ahead that may promote policy objectives in place.

5.2 Governance of Industrial Chemicals in the Baltic Sea Region over Time

Chemical risks in BSR are dealt with at local, regional, national, EU and international levels. Measures taken at the local level, being largely those of implementing law, are dependent on decisions taken at the national level and, in the case of traded products, on supranational agreements, made either within the EU or internationally. Governance structures at high levels in particular have evolved and changed substantially since the mid-1900s, the time period when industrial chemicals increasingly have come into use.

When the debate on chemical risks and their governance was initiated in the early 1960s, West Germany was the only country in BSR that was then a member of what was known as the European Community (EC). It was only a decade later or so that another Baltic Sea country, Denmark, joined the EC. At that time, the EC had no more than a rudimentary chemicals policy. What was in place in parallel was the Helsinki Convention (1974) for the protection of the Baltic Sea, which entered into force in 1980, as a binding framework agreement for the seven contracting parties. The convention aimed to control all types of pollution and imposed various obligations on parties to counteract hazardous substances. The Soviet Union was a part to the convention and dominated the eastern shores of the Baltic Sea. In spite of the cold war, however, the HELCOM operative body of the convention managed to adopt a large number of 'recommendations' over the coming decades, one example being for hazardous substances (Selin and VanDeveer 2004).

Eventually, the Berlin Wall and the Soviet Union fell, and soon thereafter, the convention was amended and other (Eastern European) countries and the EC joined the cooperation (Helsinki Convention 1992). The revised convention was extended and strengthened with specific and technical provisions and action points addressing the prevention and control of pollution including chemical pollution. The precautionary principle was explicitly included, even if the concept was referred to already in a 1988 ministerial meeting, and precautionary measures were taken already from the start, e.g. by recommending phase-out of substances not fully scientifically proven to cause damage (Pyhälä et al. 2007).

In 1995, Finland and Sweden joined the EU. Most of the countries of the Baltic Sea shoreline were then part of the EU. Due to a treaty revision in the mid-1980s (EC 1986), the EU had set explicit treaty-based environmental objectives, which over time led to a more comprehensive – and binding – environmental legislation in the union, including in the field of chemicals and the marine environment. It is important to point out that the EU has the power to enforce various stipulations if not followed by Member States, something that is not the case under the Helsinki Convention despite its binding character. In fact, the European Court of Justice can in some cases even decide to impose economic sanctions on Member States that do not comply with legislation agreed upon in common.

Ten years later in 2004 when the EU was enlarged, Russia was the only Baltic country that remained outside the EU (i.e. Poland and the three Baltic States became EU members). Since then, the EU has steadily introduced more policies that apply to the presence of chemicals in the marine environment of BSR. These include laws such as the environment-oriented Water Framework Directive and Marine Strategy Framework Directive, and the polluter-oriented REACH regulation for industrial chemicals, as well as the EU Strategy for BSR (European Commission 2009). Similar developments also took place under HELCOM, which, besides further recommendations relating to chemicals, adopted the Baltic Sea Action Plan in 2007 (HELCOM 2007). The plan has several objectives, including that life in the Baltic Sea should be 'undisturbed by hazardous substances'. It also underlines the need to apply the EAM. In the hazardous substance segment of BSAP, four 'ecological objectives' are set: concentrations of hazardous substances close to natural levels, all fish safe to eat, healthy wildlife and radioactivity at pre-Chernobyl level. These targets are then further operationalised by, for instance, various indicators.

In addition to formal governance under HELCOM and the EU, a wide array of actors and networks has strived since decades to protect the Baltic Sea environment. Among these are a number of national and international NGOs, as well as various business, city and university networks, which collaborate on marine and other governance issues.

In summary, the governance structures have shifted substantially over time in BSR both under the Helsinki Convention and within the EU, with increasingly more attention being given to an environment-oriented perspective. The geopolitical changes of the last few decades have allowed for improved collaboration and international governance structures that potentially are more capable than in the past of coping with environmental problems and risks in the Baltic Sea at an international level. In what follows, the selected three sub-cases and generations (with respect to broader societal recognition, debate and policy-making[1]) of hazardous chemicals that occur in the region, and how they are governed, will be described and analysed.

5.3 The First Generation, Chlorinated Organic Substances: PCBs

Halogenated organic substances are in general particularly problematic man-made chemicals. They are released and can be found in living organisms all around the globe, from Alaska in the North to the summits of European mountains to

[1] As will be shown, scientific studies showing or indicating problems often came earlier (see also EEA 2001, 2013).

deep-water fish in the South Atlantic (Burkow and Kalleborn 2000; Carrera et al. 2001; Looser et al. 2000). The use of chlorinated organic chemicals was questioned by scientists as far back as the 1950s (Linduska 1953) and gave rise to international public concern after the publication of Rachel Carson's 'Silent Spring' in 1962. While the substances in focus then were often pesticides designed to be toxic, for example, DDT, the widely used[2] group of polychlorinated biphenyls, PCBs, was not intended to have any toxic effects. Despite that, and long after its original introduction in 1929, scientists revealed PCB in the environment in the 1960s (Jensen 1966). This was the case because PCBs have persistent and bioaccumulative properties. These properties, in combination with the substance group's toxic properties (e.g. carcinogenicity, reproductive toxicity, environmental toxicity), led to severe adverse effects in various organisms and ecosystems. In the Baltic Sea area, for example, white-tailed eagles, grey seals, ringed seals and otters were among the species severely affected (Bergman and Olsson 1985; Helander 1983; Roos et al. 2001). Findings pertaining to new types of neurological effects on humans were reported as late as in 2001,[3] which illustrates the not seldom long time gap between the initial use of a substance and strong evidence on chronic health effects (Schantz et al. 2001).

5.3.1 EU Policy

The EU is clearly a dominant actor in BSR with regard to policy. Legislation within the EU is divided into a primary treaty level and secondary directives, regulations and decisions. The treaties set out the basis for the Council's and the Parliament's co-decisions on secondary law, as well as stipulate legal principles – in the field of the environment, for instance, precaution and that the polluter should pay. Up to the mid-1980s, laws relevant to the environment generally aimed to harmonise national legislation in order to promote the free movement of, for example, products of various kinds. Since the Single European Act (EC 1986), however, the EU has been mandated to legislate also in explicit order to protect the environment, an area where laws set minimum requirements that Member States may choose to make more stringent. If the treaty basis on the other hand is market harmonisation, it is quite difficult for a Member State to deviate from the common provisions, unless, for example, new scientific evidence shows that measures are needed to attain environmental objectives. To what extent that is possible is ultimately decided in the European Court of Justice in case of a trial in which treaty-based principles on environmental protection as well as, e.g. proportionality are considered.

When chemicals policy emerged in the European Community in the 1960s, with, for example, a directive (EEC 1967) on classification and labelling of industrial chemicals, the aim was market harmonisation. Some years later, another directive

[2] World production, for example, in the 1980s was in the order of millions of tonnes.

[3] Human toxic effects under some exposure conditions have been reported at least since the 1930s though; see EEA (2001).

(EEC 1976a) provided ground for restrictions on substances and preparations, in themselves or in products, but a ban on the use of PCBs was not included until 1985 (EEC 1985) when the EC in BSR included West Germany and Denmark. Nowadays, this ban is incorporated in the REACH regulation (EC 2006a) for industrial chemicals, which has replaced much of the earlier legislation.[4]

Besides for the general ban, a number of other policy and regulatory tools targeting PCBs have been developed, including a directive governing disposal of PCB that aims for a phase-out of equipment with PCB by 2010 (EC 1996), a 'Strategy for dioxins, furans and PCB'[5] (European Commission 2001) and recommendations and regulations for maximum levels of certain contaminants in foodstuff. Compared to directives, which Member States themselves are responsible for achieving, running the risk of ending up in the European Court of Justice in case of non-compliance, regulations are directly binding all over the EU, i.e. they have a stronger and more immediate legal power. PCBs were mentioned in the food contamination regulation of the EC (2001), but limit values came first with EC (2006b) for dioxin-like PCBs and EC (2011) for non-dioxin-like PCBs. However, Sweden, Finland and Latvia have all argued for and been granted derogations, which at the time of writing were still in place, meaning that it is allowed in these countries to sell contaminated fish to the national populations at large, in spite of opposing views from expert agencies (EC 2001; SNFA 2011).

When it comes to the presence of PCBs in the environment as such, the general 1976 directive on limit values for dangerous substances in water (EEC 1976b) did not include PCBs specifically.[6] Neither did the original Priority Substances Directive (EC 2008b), sometimes referred to as a daughter directive to the Water Framework Directive (EC 2000), which sets environmental quality standards for 33 substances or groups of substances. Recently though, amendments of the WFD and the Priority Substance Directive (PSD) included dioxin-like PCBs. In the case of PSD, Member States have to implement applicable environmental quality standards by 2018 in order to reach a good surface water chemical status by 2027 at the latest, by the means specified in the Water Framework Directive (EU 2013).

The Marine Strategy Framework Directive (EC 2008a), which is based on EAM and the precautionary principle, includes a focus on what the European Commission has decided to call 'contaminants' (according to the so-called Descriptor 8), which to a large extent are priority substances in WFD and PSD (EU 2010a), including PCBs. Member States are responsible to further define more precise targets for these hazardous substances as well as programmes to achieve a 'good environmental status' by 2020. The Baltic Sea is one region in which this has to be done (EC

[4] A general ban is also included in another EU Regulation, which aims to implement the Stockholm Convention on persistent organic pollutants, one of which is PCBs (EC 2004).

[5] The strategy lists all EU measures that by then were taken to mitigate PCB pollution (not all of these are discussed in this study), and the strategy was followed up in 2004 and 2007 (for the latter, see: http://eur-lex.europa.eu/legal-content/EN/TXT/PDF/?uri=CELEX:52007DC0396&from=EN).

[6] Organohalogens were referred to in general though.

2008a). When it comes to the Baltic Sea, HELCOM has established a link between MSFD and the Baltic Sea Action Plan and coordinates the national implementation by EU Member States (HELCOM 2010b).

5.3.2 HELCOM Policy

Under the 1974 Helsinki Convention, organochlorinated and several other hazardous substances became targets for HELCOM's activities. Since then, PCBs have been in focus for various monitoring (in a coordinated manner since 1979), risk assessment and risk management measures. Monitoring and assessments under HELCOM are following a specific strategy with, for example, objectives, principles and indicators (HELCOM 2013a), and the contracting parties as well as various expert groups and scientific committees participate in the work. A set of PCBs belonging to the so-called HELCOM 'core indicators' for hazardous substances under the Baltic Sea Action Plan (HELCOM 2007, 2013), and their trends are monitored over time (Boalt et al. 2013).

In terms of risk management, HELCOM measures commonly consist of various Recommendations to the parties of the convention. Regarding PCBs, the 1974 Convention mentioned them explicitly. Based on a 1981 ICES environmental assessment, HELCOM adopted a Recommendation (3/1) in 1982 that contracting parties should, allowing for some exemptions, prohibit the introduction of new products containing PCBs and develop national regulations for reducing discharges from existing sources (HELCOM 1982). In 1985, another Recommendation (6/1) underlined the urgency to do so and requested the parties to stop production and marketing of PCBs from 1987, more specific action against existing sources as well as the use of more specified reporting procedures (HELCOM 1985).

Due to the discovery of elevated concentrations of numerous hazardous substances in the Baltic Sea in the 1980s, a 1988 Ministerial Declaration took a broad approach and set a target to reduce the total discharges of the most harmful substances by around 50 % by 1995 (HELCOM 1988). This led to a list of such substances in 1991, initially covering 46 substances and groups of substances, 1 being the large group of organohalogens (Selin and VanDeveer 2004).

Similarly, the 1992 amendment of the Convention required the Contracting Parties to take a broad approach in order 'to prevent and eliminate pollution of the marine environment of the Baltic Sea Area caused by harmful substances from all sources' (Helsinki Convention 1992, Article 5). Annex 1 contained general principles for 'harmful substances' and listed both priority groups for action and 'banned' substances, where it was stated (Part 2.2) that in 'order to protect the Baltic Sea Area PCBs shall be banned for all uses, except in existing closed system equipment until the end of service life or for research, development and analytical purposes in the Baltic Sea Area and its catchment area.'

Furthermore, based on a 1996 agreement taken in the Council of the Baltic Sea States (CBSS 1996), another HELCOM Recommendation (19/5) in 1998 set a target to continuously reduce discharges, emissions and losses of hazardous substances towards cessation by 2020, so as to reach background values for naturally occurring substances and close to zero concentrations for man-made substances (HELCOM 1998). By then, 280 substances were listed as 'potential substances of concern', of which 43 were prioritised for 'immediate action', including PCBs. Compared to previous agreements that focused on upstream measures, the starting point here was environmental quality and close to zero tolerance for pollution.

The PCB problems remained in the 2000s, and a new Recommendation (25/1) on PCBs was adopted, superseding a previous one (6/1), with updated calls on the Parties and further specifications on measures relating to, for example, destruction and decontamination (HELCOM 2004). In the 2007 Baltic Sea Action Plan, 11 of the most problematic substances, including dioxin-like PCBs, were listed as being of 'specific concern' to the Baltic Sea, and for these, ecosystem-based targets have been developed in order to reach the objective of a 'life undisturbed by hazardous substances'.

Finally, also the broad 1998 Recommendation (19/5) was replaced in 2010 with a new general Recommendation (31E/1) on 'Implementing HELCOM's objectives for hazardous substances' (HELCOM 2010c). Here, the attached guiding principles, definitions and strategy were updated and modernised. For instance, there are explicit references to both the EU REACH Regulation and the UN Global Harmonised System for classification and labelling of chemicals. The precautionary elements are more obvious, including a definition (and linked to priority setting and mitigating measures) of 'hazardous' to include substances that are very persistent and very bioaccumulative (i.e. without necessarily being toxic). PCBs are among the listed priority substances.

5.3.3 Analysis of Measures Taken and Approaches Applied

Due to the various regulatory initiatives, PCB concentrations in the Baltic Sea marine environment have decreased substantially over time, even though they have stabilised at levels that still are significantly above those of pre-industrial era. The populations of some of the wildlife species that previously were seriously threatened such as the three Baltic Sea seal species and the white-tailed eagle have consequently increased.[7] However, according to HELCOM, the general situation is still considered as 'moderate or bad' (Boalt et al. 2013), and the objective to reach an environment with levels of man-made substances close to zero is far from achieved, also with respect to PCBs. For example, the 1988 goal to half discharges of the most harmful substances by 1995 had by 2001 only been achieved for less than two-thirds of the substances on the list. However, the target had been achieved for PCBs (Selin and VanDeveer 2004). Moreover, despite precautionary measures taken, the total remediation and waste management costs for PCBs continue to be very high. In

[7] In the case of the white-tailed eagle, artificial feeding for three to four decades played a key role in preventing national or regional extinction.

Sweden alone, from 1971 to 2018, the costs were estimated to be 380–480 million euros. Corresponding EU figures amounted to 15–75 billion euros (NCM 2004).

Looking back, it is clear that the legally stipulated practical measures against PCB in the 1970s in some Baltic Sea countries were taken years before there was comprehensive conclusive evidence on causal links between PCB contamination and the various adverse effects observed in the Baltic Sea environment. In spite of some but much earlier evidence on certain negative health effects of PCB exposure (EEA 2001), it can therefore still be said that these policies were precautionary.

Moreover, it is clear that HELCOM has been an important body for much of the policy-making with regard to PCBs. However, forerunner countries (e.g. Sweden and Germany[8]) acted earlier and to some extent independently of HELCOM. But laggard countries (e.g. Poland) did not act until they were applicants to or members of the EU. The forerunners have foremost used HELCOM to push for measures in other countries, by insisting on and making use of both recommendations and associated monitoring and risk assessment activities. Conversely, HELCOM has hardly played a decisive role for chemicals policy in laggard countries, for instance, in Poland, which seemingly (although it is important to note that data gaps for Russia give rise to significant uncertainty) stands for most of the PCB emissions (COHIBA 2012), and where the societal debate on chemicals issues is largely absent (Eriksson et al. 2010a).

Compared with HELCOM, policy measures in the EU came later, but the impact of EU measures on national policy was significant, including in laggard countries. Much therefore speaks for the regulatory power of the EU to ultimately have been of higher importance for the abatement in practise of PCB problems, than what the power of HELCOM has been. However, HELCOM has been instrumental in acting early and setting the agenda, in monitoring and assessing environmental quality and in showing – at least initially – through its various recommendations the importance of the regulatory way forwards. It is far from sure that the EU would have acted as it did without this pioneering, catalytic and facilitating role of HELCOM. In addition, HELCOM in contrast to the EU includes Russia, where a number of the otherwise restricted hazardous substances are still permitted (COHIBA 2012).

Moreover, HELCOM has taken a leading role in developing environment-oriented approaches. This has been the case with the 1992 Helsinki Convention, the zero concentration objective adopted in a Recommendation in 1998 and in the joint HELCOM-OSPAR EAM-statement in 2003. Similarly, the precautionary principle has definitely and for a longer time been playing a more central role in HELCOM than in the EU. More recently, the EU has also institutionalised EAM, for example, in MSFD. Considering the links and increasing coordination of implementation between MSFD and BSAP, it seems plausible that both institutions will strive for a broader use of EAM, even if the EU at the same time will keep its strongly polluter-oriented REACH regulation.

All in all, it seems well motivated to conclude that in this PCB sub-case, HELCOM and the EU have interplayed in a positive way, the former mainly initiat-

[8] In, e.g. Germany, Monsanto and Bayer stopped PCB production by 1977 and 1983, respectively, which might have facilitated some of the regulation.

ing measures and being an agenda-setter and conceptual pioneer and the latter as a powerful and adaptive legislator. It must be underlined though that this relates to policies for PCB, one of the most infamous pollutants in the world, and generalisations cannot be made to effect and effectiveness for chemicals policy at large. On the contrary, it can be concluded that the regulatory response even in this notorious case has been far from rapid.

5.4 The Second Generation, Brominated Organic Substances: PBDEs

A second group of industrial organohalogens is the brominated, and within this group, not least flame retardants and particularly polybrominated diphenyl ethers (PBDEs) have been much discussed since at least the 1980s. Brominated flame retardants (BFRs) have been frequently used since the 1950s as additives in plastic polymers in, for example, textiles and electric and electronic products. Since PBDEs generally are persistent and bioaccumulative, they have increasingly been detected in wildlife, humans and the wider global environment since the 1980s (Birnbaum and Staskal 2004; Law et al. 2006; Nyberg et al. 2013a; de Wit 2002). There are still knowledge gaps when it comes to the specific properties and risks of most BFRs. However, so-called lower brominated[9] PBDEs have since long been shown widespread in ecosystems, including in the Baltic Sea environment (Airaksinen et al. 2014; Szlinder-Richert et al. 2010), and are known to negatively affect hormone, reproductive and neurological systems in mammals (Birnbaum and Staskal 2004; Darnerud et al. 2001; Eriksson et al. 2001, 2002; Ilonka et al. 2000). More recent studies have shown great concerns regarding human health effects as well (Eskenazi et al. 2013; Herbstman et al. 2010; Ward et al. 2014). Higher brominated PBDEs, such as decaBDE, were initially not considered as problematic, and many industry stakeholders in fact claimed that decaBDE did not bioaccumulate due to the molecule's comparatively large size (ECB 2002). But this was falsified when Lindberg et al. (2004) showed the presence of decaBDE in eggs of the peregrine falcon. Other studies have pointed out that decaBDE can degrade in the environment to lower brominated PBDEs (Gerecke et al. 2005; Stapleton et al. 2004).

5.4.1 EU Policy

Scientific findings from the 1980s and onwards did not lead to any potential risk mitigating measures in the EU until the mid-1990s when some BFRs were placed on so-called priority lists for risk assessment of existing substances (EEC 1993).

[9] PBDEs with not more than five bromine atoms, such as pentaBDE.

Among these were pentaBDE and octaBDE. The initial risk assessment for pentaBDE recognised the need for risk reduction measures (ECB 2000); additional studies highlighted further problems, as exemplified above. Based on the risk assessment, the EU in 2004 eventually prohibited use of not only pentaBDE but also octaBDE (EC 2003a). The additional regulation of decaBDE was intensively debated, but the outcome of the European Parliament and Council co-decision procedure, despite that the former body wanted a ban, was that the substance was not banned.[10] In spite of that, political and legal processes concerning implementation of the ROHS directive on electric and electronic products (EC 2003b), which banned pentaBDE and octaBDE from 2006, led to a 2008 ban of decaBDE as well (ECJ 2008; Eriksson et al. 2010a, 2010b).[11]

In parallel, the 2003 WEEE directive (regulating waste from electrical and electronic equipment) set recycling and reuse targets which were relevant in conjunction to some plastics with brominated flame retardants (EC 2002).

More recently, decaBDE has been registered under the REACH regulation (in September 2010, based on the data from the previous EU risk assessment) and in 2012 it was placed on the so-called Candidate List, as a Substance of Very High Concern (SVHC). It thereby became a substance in need of potential authorisation and restriction,[12] which in 2013 led the European Chemicals Agency to start preparing a restriction proposal (ECHA 2014).

Turning to environmental quality and the Water Framework Directive, the linked Priority Substance Directive originally set various environmental quality standards for surface water for 33 substances or groups of substances, which Member States were obliged to comply with as a main rule by 2015 (EC 2008b) and partly through means specified in the WFD. Among these substances, pentaBDE belonged to a subgroup of particular concern referred to as ‘priority hazardous substances’, whereas octaBDE and decaBDE were considered ‘priority substances’. In the recent amendment of the directive, standards for PBDEs were both amended and widened to include biota, for which the requirements must be met by 2021 at the latest (EU 2013).

Just as for PCBs, the Marine Strategy Framework Directive (EC 2008a) includes descriptors that cover PBDEs, and based on monitoring programmes and a programme of measures to be implemented by 2015–2016, the overall target of ‘good environmental status’ is supposed to be reached by 2020. In the context of the Baltic Sea, this work is coordinated by HELCOM and linked to the implementation of BSAP.

---

[10] Not restricting decaBDE was based on the recommendation of the EU 2002 risk assessment report, written by France and the UK, which stated environmental risks to be acceptable, partly based on the view that decaBDE did not bioaccumulate, even though the report recommended monitoring and further health-related studies. In addition, industry committed to reduce emissions. When the report was updated with the same general conclusion in 2004, it was criticised by the Commission’s advisory Scientific Committee on Health and Environmental Risks (see further in SCHER 2005; Eriksson et al. 2010b).

[11] Since then, the ROHS directive has been further amended.

[12] UK submitted an Annex XV Dossier to ECHA proposing decaBDE as a SVHC.

5.4.2 HELCOM Policy

By the time brominated flame retardants started to be targeted by policymakers, the Helsinki Convention had been amended, requiring the Parties 'to prevent and eliminate pollution of the marine environment of the Baltic Sea Area caused by harmful substances from all sources'. Compared to the 1974 convention, the 1992 version is more stringent and comprehensive. One example is that the 1974 Annex II list of 'noxious substances and materials' has been broadened to a 1992 Annex I list of 'Priority groups of harmful substances', where, for example, (all) 'organohalogen compounds' are included, as opposed to only 'persistent halogenated hydrocarbons' in 1974, meaning that now all PBDEs are definitely covered at least indirectly.

In the mentioned 1998 Recommendation (19/5) on the 2020 target, brominated flame retardants were included in the list of 280 substances of potential concern, but neither the BFR-group as a whole nor organohalogens and PBDEs were among the 42 substances prioritised for immediate action. However, explicit reference to PBDEs was repeatedly made in the 2007 Baltic Sea Action Plan (HELCOM 2007), which updated the Recommendation. PentaBDE and octaBDE were targeted for use, production and marketing bans by 2010, whereas decaBDE could be a target from 2009 for less stringent measures, in some sectors, if further assessments showed a need for taking such action. All three PBDEs were included in a group of 11 substances of 'specific concern' to the Baltic Sea.

Moreover, 3 years later, when the 1990 Recommendation (19/5) was updated, all three – penta, octa and decaPBDE – were included in the list of priority hazardous substances, against which measures should be focused (HELCOM 2010c).

5.4.3 Analysis of Measures Taken and Approaches Applied

According to HELCOM, the PBDE levels in, for example, fish and guillemots are generally higher in almost all monitored areas in the Baltic Sea than what is defined to be a Good Environmental Status (Nyberg et al. 2013a). At the same time, the concentrations of some individual PBDEs are decreasing, which has been claimed to be a result of the EU restrictions in place since 2004, even though data is missing for some marine areas (Nyberg et al. 2013a). It thus seems plausible that the legal measures against pentaBDE and octaBDE have given results. However, since many products with PBDEs remain in use and since the restrictions have not targeted decaBDE, which as such is a problematic substance and in addition can be degraded in the environment into, for example, pentaBDE, the environment is still affected by this group of brominated flame retardants. It is moreover evident that it took a long time for measures to be implemented, in particular those for decaBDE. Even if the time between substance introduction and policy-making is far from as long in total as for PCBs, the general awareness of chemical risks – and the science-policy arsenal available – was much stronger after the 1980s than in the 1960s, so it could well be argued that society responded comparatively slower on PBDEs than on PCBs. When measures were agreed though, the regulatory strength of the EU was much greater than in the 1960s, even though the complex science-policy interface in EU legislation, placing a very strong burden of proof for measures, led to significant policy delays.[13]

When it comes to the policy approach, the PBDE group has been regulated mainly on the basis of the chemicals and polluter-oriented policy, not based on an environmental approach, and even less so on a precautionary approach. The latter is particularly true for decaBDE, where action was stalled during lengthy and politicised decision-making and legal processes (Eriksson et al. 2010a, 2010b). The environment-oriented approaches in the case of MSFD, for example, have so far not led to significant phase-out measures in themselves, and there are obvious implementation problems. In the case of WFD, in its third implementation report, the European Commission was unable to even establish a baseline for the chemical status of surface waters due to data gaps and insufficient monitoring (more than 40 % of surface water bodies in the EU were reported as having 'unknown chemical status') making it very difficult to estimate the EU-wide or Baltic Sea situation (European Commission 2012). Similarly, while BSAP is based on EAM (in theory linking hazardous substances, negatively, to biodiversity objectives on the one hand and to eutrophication on the other, where goal achievement may lead to higher concentrations of pollutants in biota), the practical responses to these insights still have to be implemented broadly. Nevertheless, BSAP sets out to link the work of HELCOM-identified substances to the national implementation plans under WFD, thereby helping policy efficiency in practice. More generally, the implementation of BSAP may help to broaden the implementation of WFD in EU Member States, and conversely, some environment-oriented EU policies may then be implemented in the HELCOM context. As for PCBs, there is thus a positive interplay between EU and HELCOM policies, when it comes to BFRs.

5.5 The Third Generation, Perfluorinated Organic Substances: PFOS and PFOA

Man-made fluorinated organic chemicals are since long widely used in industrial processes and commercial products and are commonly detected in organisms and various environmental compartments, the most well-known case being the infamous chlorofluorocarbons that have depleted the ozone layer (e.g. EEA 2001). One complex group of fluorinated organohalogens consists of perfluorinated substances, which, due to not least their stability and surface-active properties, are commonly used since the 1950s in, for instance, firefighting foam, non-stick coatings, food packaging and electronics, as well as for water and stain proofing in textiles and

[13] Comprehensive policies for destruction and decontamination of products containing PBDEs, as exist for PCBs, are still largely missing for PBDEs (see, e.g. Bergman 2012).

shoes, used both as sprays and as components in garments and leather (Giesy and Kannan 2002; Lindstrom et al. 2011). While not all perfluorinated substances are necessarily problematic, some are indeed so. For instance, PFOS (perfluorooctane sulphonate) and PFOA (perfluorooctanoic acid) have at least since 2000s been targets of public policy (Renner 2001).

Several perfluorinated substances belonging to the group of perfluoroalkylated substances (PFAS) are persistent, bioaccumulative and toxic (Scheringer et al. 2014). PFOS and PFOA belong to the PFAS group, are both substances in use in themselves and common degradation products of hundreds of other perfluorinated substances and are extremely persistent and therefore detected around the globe, including in the Arctic environment, in Europe and the Baltic Sea (Butt et al. 2010; Holmström et al. 2005; Lau et al. 2007; Letcher et al. 2010; Nyberg et al. 2013b; Pistocchi and Loos 2009).

PFOS meets the criteria in the EU REACH regulation of being very persistent and at least bioaccumulative (RPA 2004). It also biomagnifies and is commonly found in, for instance, polar bears in the Arctic (Greaves and Letcher 2013) as well as in fish and seals in the Baltic Sea (Kratzer et al. 2011; Schuetze et al. 2010). PFOA is also very persistent and bioaccumulates in at least air-breathing mammals, including humans (ECHA 2013),[14] the first findings of such bioaccumulation being reported as early as the 1970s (Lindstrom et al. 2011). Even if there is vast uncertainty regarding the properties and effects of most substances in the PFAS group, PFOS has been classified as toxic, and animal experiments have shown negative effects of it on, for example, reproduction, endocrine systems, the liver, the immune system and the nervous system (Austin et al. 2003; Johansson et al. 2008; Lau et al. 2007). For birds of prey, observed concentrations may be close to those where effects in the environment might be seen (Nyberg et al. 2013b). Humans are exposed via water and food, such as fish (Borg and Håkansson 2012), and various hormonal and reproductive effects and risks on humans have been shown (Lopez-Espinosa et al. 2011; Lopez-Espinosa et al. 2012). PFOA has also been classified as toxic from different points of view, among these as a suspected carcinogen and toxic for reproduction (UNEP 2006; ECHA 2013). Recent studies show even more worrying signs regarding cancer (Barry et al. 2013).

5.5.1 EU Policy

In the EU, a process to restrict PFOS was started in 2005 by the European Commission, based on an OECD hazard assessment as well as the EU risk assessment report under the previous existing substance programme (European Commission 2005). In the following co-decision procedure, the European Parliament proposed, agreeing across party lines, to extend the Commission's proposal to

[14] Strangely, in spite of this judgment from ECHA, the researchers reporting on HELCOM core indicators did not consider PFOA to be bioaccumulative (Nyberg et al. 2013b).

restrict PFOS to include PFOA as well. This proposal was motivated with a reference saying that the US EPA had found the risks of the latter substance to be of 'similar concern' (European Parliament 2006). After an unusually rapid legislative process, the Parliament and Council then agreed on a directive that was more restrictive than the Commission's proposal on general limit values for PFOS in products and against firefighting foams, but did not regulate PFOA, as the Parliament had proposed (EC 2006c). The restriction was eventually moved over to the REACH regulation when it entered into force and then to EU's so-called POPs regulation (EU 2010b).[15]

PFOA, as mentioned above, was not restricted in the EU co-decision procedure, but it was said that the European Commission shall keep substitutes and ongoing risk assessment activities under review and propose risk-reducing measures when needed (EC 2006c). Presently, PFOA is on the REACH Candidate List as a Substance of Very High Concern, which means that it eventually might be a target for an authorisation process (ECHA 2013).

Turning to environmental quality, PFOS was included in WFD and the Priority Substance Directive after it was revised recently. The Priority Substance Directive required Member States to implement stated quality standards by 2018 and attain a good surface water chemical status by 2027 at the latest (EU 2013). The quality standard is based on the most sensitive parameter for PFOS, namely, secondary poisoning (KEMI 2013b). In MSFD, one of the descriptors (number 8) covers PFOS, and on that basis, Member States shall define precise targets for a 'good environmental status' to reach by 2020 at the latest, through programmes to be implemented in 2015–2016 at the latest (EC 2008a). No similar environmental quality stipulations exist for PFOA.

5.5.2 HELCOM Policy

PFOS was included in the work of HELCOM in particular after BSAP was adopted in 2007. In the plan, the Parties agreed to 'start by 2008 to work for strict restrictions on the use in the whole Baltic Sea catchment area of the Contracting States of… perfluorooctane sulfonate (PFOS)'. PFOA was also included in BSAP in the same manner as decaBDE, namely, that by 2009, 'if relevant assessments show the need' to initiate adequate measures in some sectors, for instance, use restrictions (HELCOM 2007). PFOS and PFOA were both listed among the 11 substances of 'specific concern' in BSAP and in the most recent Recommendation (31E/1) listing 'priority hazardous substances' (HELCOM 2010c), but only PFOS later became a HELCOM Core indicator (HELCOM 2013b).

---

[15] PentaBDE was included at the same time. In simultaneous amendment of Annex IV and V of the same regulation (http://eur-lex.europa.eu/LexUriServ/LexUriServ.do?uri=OJ:L:2010:223:0020:0028:EN:PDF), new provisions regarding waste management also came.

5.5.3 Analysis of Measures Taken and Approaches Applied

When it comes to PFOS, the concentrations in the Baltic Sea have increased since the 1960s, in some case exponentially (Holmström et al. 2005). Even if there are a few recent signs of decline, PFOS levels exceed the thresholds in many monitoring sites, 'indicating moderate or even bad environmental status', with the highest PFOS levels in biota found in top consumers such as grey, harbour and ringed seals (Nyberg et al. 2013b). Much data on temporal trends is missing, and to what extent PFOS levels really have started to decline in, for example, the Baltic Sea is therefore far from certain. Decreased levels in human blood have, however, been reported in the USA in the 2000s, allegedly following the dominating producer 3 M-Corporations voluntary measure started in 2000 to phase out PFOS and related chemicals (Renner 2008). Less is known about temporal development of PFOA in the Baltic Sea, and trends can of course point in different directions depending on if emissions, water concentrations or levels in biota are studied, with variations between regions and species. Nevertheless, despite a downturn in emissions, also due to a voluntary phase-out by large producers ('the PFOA Stewardship programme'), levels of PFOA in the Arctic sea water have been predicted to continue to increase until around 2030, whereas the situation has been predicted to improve in the Northern Temperate zone in the 2010s (Butt et al. 2010).

From a regulatory point of view, the legislative process to restrict PFOS in the EU was comparatively rapid, even if it – as so often otherwise – was initiated long after science indicated problems existed. Here, the EU has been the dominating phase-out force, whereas HELCOM has focused more on principal policy advice and assessment, seemingly in the aftermath of the EU legislative process (whereas HELCOM policy and assessments were ahead of legislation when it comes to PCBs). Moreover, regulatory action took place in particular in the sphere of chemicals policy and was not based on environment-oriented legislation. Partly, the rapid regulatory process can of course be attributed to the fact that many industries had beforehand already promised a voluntary phase-out. Given that production-related emissions for these reasons will cease or at least continue to decline, it is difficult to predict which roles MSFD and PSD will play in this particular case in the future. Perhaps, MSFD's focus on environmental quality might speed up the upstream work so that increased focus on disposal and sanitation might follow, in order to try to cope with still existing products in use.

Public policy has not focused on PFOA as it has on PFOS, and despite science-based identification of problems with PFOA, restrictions are still not in place. Here, HELCOM has taken a more pioneering role in terms of policy direction, but it remains to be seen what that will lead to; HELCOM parties are not taking country-based measures to the same extent after EU enlargement and the REACH regulation as they (at least some of them) did before.[16] In many ways, PFOA seems to have

[16] Some EU Member States' initiatives regarding BPA and decaBDE show that national measures are not completely impossible in an EU harmonised policy arena.

been considered in a similar way to decaBDE in the case of PBDEs, i.e. showing potentially less worrying properties and therefore being given, at most, secondary attention in the policy process. From a broader point of view, it also remains to be seen to what extent the use of other perfluorinated substances with similar hazardous properties will increase, following the phase-out of PFOS and, potentially, PFOA. Science-based concerns are not necessarily smaller in those cases (Scheringer et al. 2014).

5.6 Discussion

This study has focused on how public risk governance of three generations (chlorinated, brominated, fluorinated) of hazardous chemicals and specific examples linked to these (PCBs, PBDEs, PFOS/PFOA) has developed in BSR over time, with a special focus on the EU and HELCOM.

Indeed, both the EU and HELCOM have responded with various types of legislation, recommendations and policies in order to manage the problems and risks caused by these substances. In general, the response has been more reactive than proactive, but when the stricter types of measures pointed out in the article once have been taken, both problems and risks have decreased over time, even though not to the extent needed in order to reach the overall objectives of a non-toxic environment.

If we look at the question of time, the regulatory response to societal debate might generally be seen as more rapid (or less slow) in recent years, than in the past, both in the EU and under the Helsinki Convention. Even if the fluorinated substances (third generation) have been in use for a long time, the EU and HELCOM have reacted more firmly in the PFOS case than in the PCB case. That is what could be expected given the more solid knowledge base and the higher environmental awareness today.

Still, given this improved state of knowledge and awareness on chemical risks, and given decades of recurring experiences of regulatory bottlenecks, one must ask if the decision-making processes in the EU are not unreasonably slow today. The question must even be raised if these processes might actually be *comparatively* slower than in the past. For example, the management of decaBDE (second generation) has been characterised by significantly stronger requirements than in the PCB case (first generation), to produce an overwhelming body of evidence, based on quite traditional risk assessment processes, despite the fact that history has shown this to be problematic and despite the fact that precaution nowadays is a part of the EU treaty as well as secondary law.[17] Further scholarly studies on these and other substance cases are needed though, to be able to draw firm conclusions on the topic,

---

[17] The Treaty on the Functioning of the EU (TFEU 2007) and REACH (e.g. authorisation can be demanded for hazardous substances even without necessarily proving them to be toxic) include precautionary elements.

including on whether the precautionary dimension that after all exists in REACH is counteracted by complex and slow implementation procedures so that the regulatory speed is even slower today than under the previous EU risk assessment programme for existing substances. A corresponding question needs to be asked also concerning HELCOM, which – despite more commonly than EU referring to the precautionary principle – has been quite reluctant to act on decaBDE and PFOA, albeit not to the same extent as the EU.

All in all, this regulatory reluctance signals a weak tradition and capacity in both institutions to cope with uncertainty, which may fuel sociopolitical controversy. For example, in the case of decaBDE assessment and regulation in the EU, divergent opinions on how to interpret available knowledge and remaining uncertainty opened the way for politicisation of the issue and consequently controversial regulatory and court processes. Without policy and regulatory reforms, which most importantly need to move towards a more fully reversed burden of proof,[18] substantial difficulties will remain even in the future when managing hazardous chemicals and their environmental and health risks, thereby jeopardising the agreed objectives in e.g. BSAP.[19]

Turning from the dominating risk-based and polluter-oriented chemical regulations to the more environment-oriented directives and plans, such as MSFD and BSAP, they clearly stand for a more holistic perspective, being in line with EAM, often expressing precaution as important. From the aquatic starting point, these tools aim at addressing a number of substances based on environmental monitoring and stated limit values. Here as well, however, decaBDE and PFOA have been included at quite a late stage. Moreover, even if identified as being of concern, MSFD and BSAP as such do not lead to phasing out or restrictions on substances that are targeted. In addition, country-based implementation is far from effective. Despite observed regulatory hurdles such as the high burden of proof in EU chemicals policy, it is therefore still not certain that environment-oriented policies such as MSFD and BSAP are more efficient in promoting the agreed objectives of, for example, good environmental status.

Based on the reasoning above, it seems obvious that both policy orientations and approaches are needed and that they need to be better linked than today. Our case studies indicate that this coordination is required both in science (e.g. in terms of information and knowledge exchange between REACH and MSFD/BSAP) and management (e.g. by allowing fast-tracking in REACH of hazardous substances that show up in the marine environment or by triggering upstream sanitation measures if prohibited hazardous substances continue to show up in the environment as result of leakage from already introduced products in society). In addition, starting from EAM, both MSFD and BSAP should reasonably be developed to include mechanisms for addressing groups of similar hazardous substances and other types

[18] For example, the reversed burden of proof in REACH concerns substance registration but not, for example, restrictions (see further in Karlsson 2010).

[19] See, e.g. Karlsson et al. (2011) and Karlsson (2010) for more detailed ideas on governance reforms.

of problematic mixtures. In particular, when evidence of continuing problems with regard to halogenated substances is as strong as it is, illustrated in this study with a few examples, the EU and HELCOM need to implement chemical group perspectives, if the objective of a non-toxic environment is to be reached. Science gives stronger support to the default assumption that similar types of substances within a group – let it be chlorinated, brominated or fluorinated, as well as organohalogens – show significantly similar properties than to the default assumption that these substances would have substantially different properties. A broader group-based perspective in chemicals and environmental policy would also allow for much more rapid regulatory processes, promoting the overall objectives in place.

Finally, looking ahead, it is clear that the EU over time has emerged as the strongest policy-making body in the governance of hazardous chemicals in BSR. However, HELCOM still plays an important role for monitoring and environmentally based assessments, as well as for the conceptual framing of measures and appropriate strategies. The EU and HELCOM should be seen to be complementary, more or less well coordinated in different parts of the region, rather than as competing institutions.[20] Europeanisation and regionalisation can be synergetic trends, given the right set-up.

Acknowledgements This research was funded by the Foundation for Baltic and East European Studies and the European Community's Seventh Framework Programme (2007–2013) under grant agreement n° 217246 made with the joint Baltic Sea research and development programme BONUS, as well as by the Swedish Environmental Protection Agency and the Swedish Research Council FORMAS. We wish to thank these institutions for enabling this research. Two peer reviewers are also thanked for valuable comments on an earlier version of the chapter.

References

Adger WN, Jordan A (eds) (2009) Governing sustainability. Cambridge University Press, Cambridge

Airaksinen R, Hallikainen A, Rantakokko P, Ruokojärvi P, Vuorinen PJ, Parmanne R, Verta M, Mannio J, Kiviranta H (2014) Time trends and congener profiles of PCDD/Fs, PCBs, and

[20] In addition, HELCOM includes, e.g. Russia, which is evidently important to be included in the cooperation in order to promote the targets set up.

PBDEs in Baltic herring off the coast of Finland during 1978–2009. Chemosphere 114:165–171
Allanou R, Hansen BG, van der Bilt Y (1999) Public availability of data on EU high production volume chemicals. European Commission, European Chemicals Bureau, Ispra
Austin ME, Badrinarayanan SK, Barber M, Kurunthachalam K, MohanKumar PS, MohanKumar SMJ (2003) Neuroendocrine effects of perfluorooctane sulfonate in rats. Environ Health Perspect 111:1485–1489
Backer H, Leppänen JM, Brusendorff AC, Forsius K, Stankiewicz M, Mehtonen J, Pyhälä M, Laamanen M et al (2010) HELCOM Baltic Sea Action Plan. A regional programme of measures for the marine environment based on the ecosystem approach. Mar Pollut Bull 60:642–649
Barry V, Winquist A, Steenland K (2013) Perfluorooctanoic acid (PFOA) exposures and incident cancers among adults living near a chemical plant. Environ Health Perspect 121:1313–1318
Bergman A (2007) Pathological changes in seals in Swedish waters: the relation to environmental pollution. Doctoral thesis, Swedish University of Agricultural Sciences, Uppsala
Bergman P (2012) Bättre EU-regler för en giftfri miljö (in Swedish). Rapport 1/12. KEMI, Stockholm
Bergman A, Olsson M (1985) Pathology of Baltic grey seal and ringed seal females with special reference to adrenocortical hyperplasia: is environmental pollution the cause of a widely distributed disease syndrome? Finn Game Res 44:47–62
Birnbaum LS, Staskal DF (2004) Brominated flame retardants: cause for concern? Environ Health Perspect 112:9–17
Boalt E, Nyberg E, Bignert A, Hedman J, Danielson S, the HELCOM CORESET expert group for hazardous substances (2013) Polychlorinated biphenyls (PCB) and dioxins and furans. HELCOM, Helsinki
Borg D, Håkansson H (2012) Environmental and health risk assessment of perfluoroalkylated and polyfluoroalkylated substances (PFASs) in Sweden. Report 6513. SEPA, Stockholm
Burkow IC, Kalleborn R (2000) Sources and transport of persistent pollutants to the arctic. Toxicol Lett 112–113:97–92
Butt CM, Berger U, Bossi R, Tomy GT (2010) Levels and trends of poly- and perfluorinated compounds in the arctic environment. Sci Total Environ 408:2936–2965
Carrera G, Fernandez P, Vilanova RM, Grimalt JO (2001) Persistent organic pollutants in snow from European high mountain areas. Atmos Environ 35:245–254
CBSS (1996) Kalmar Communiqué Council of the Baltic Sea States. Fifth Ministerial Session, Kalmar 2–3 July 1996
COHIBA (2012) Major sources and flows of the Baltic Sea Action Plan hazardous substances. WP4 final report. IVL, Stockholm
Curtin R, Prellezo R (2010) Understanding marine ecosystem based management: a literature review. Mar Policy 34:821–830
Darnerud PO, Eriksen GS, Johannesson T, Larsen PB, Viluksela M (2001) Polybrominated diphenyl ethers: occurrence, dietary exposure, and toxicology. Environ Health Perspect 109:49–68
de Wit CA (2002) An overview of brominated flame retardants in the environment. Chemosphere 46:583–624
EC (1986) Single European Act. OJ L 169:1–29
EC (1996) Council Directive 96/59/EC on the disposal of polychlorinated biphenyls and polychlorinated terphenyls (PCB/PCT). OJ L 243:31–35
EC (2000) Directive 2000/60/EC of the European Parliament and of the Council establishing a framework for community action in the field of water policy. OJ L 327:1–72
EC (2001) Council Regulation 2375/2001/EC of 29 November 2001 amending Commission Regulation 466/2001/EC setting maximum levels for certain contaminants in foodstuffs. OJ L 321:1–5
EC (2002) Directive 2002/96/EC of the European Parliament and of the Council on waste electrical and electronic equipment (WEEE). OJ L 37:24–39

EC (2003a) Directive 2003/11/EC of the European Parliament and of the Council amending for the 24th time Council Directive 76/769/EEC relating to restrictions of certain dangerous substances and preparations (pentabromodiphenyl ether, octabromodiphenyl ether). OJ L 42:45–46

EC (2003b) Directive 2002/95/EC of the European Parliament and of the Council on the restriction of the use of certain hazardous substances in electrical and electronic equipment. OJ L37:19–23

EC (2004) Regulation (EC) No 850/2004 of the European Parliament and of the Council of 29 April 2004 on persistent organic pollutants and amending Directive 79/117/EEC. OJ L 158:7–49

EC (2006a) Regulation (EC) 1907/2006 of the European Parliament and of the Council concerning the Registration, Evaluation, Authorisation and Restriction of Chemicals (REACH). OJ L 396:1–849

EC (2006b) Commission Regulation 1881/2006/EC setting maximum levels for certain contaminants in foodstuffs. OJ L 364:5–24

EC (2006c) Directive 2006/122/EC of the European Parliament and of the Council amending for the 30th time Council Directive 76/769/EEC on restrictions of certain dangerous substances and preparations (perfluorooctane sulfonates). OJ L 372:32–34

EC (2008a) Directive 2008/56/EC of the European Parliament and of the Council establishing a framework for community action in the field of marine environmental policy (Marine Strategy Framework Directive). OJ L 164:19–40

EC (2008b) Directive 2008/105/EC of the European Parliament and of the Council on environmental quality standards in the field of water policy. OJ L 348:84–97

EC (2011) Commission Regulation 1259/2011/EU of 2 December 2011 amending Regulation EC 1881/2006 as regards maximum levels for dioxins, dioxin-like PCBs and non dioxin-like PCBs in foodstuffs. OJ L 320:18–23

ECB (European Chemicals Bureau) (2000) European Union risk assessment report diphenyl ether, pentabromo deriv. Institute for Health and Consumer Protection, Joint Research Centre, European Commission

ECB (European Chemicals Bureau) (2002) European Union risk assessment report bis(pentabromodiphenyl) ether. Institute for Health and Consumer Protection, Joint Research Centre, European Commission

ECHA (European Chemicals Agency) (2013) Member State Committee support document for identification of pentadecafluorooctanoic acid (PFOA) as a substance of very high concern because of its CMR and PBT properties. ECHA, Helsinki

ECHA (European Chemicals Agency) (2014) Annex XV restriction report. Proposal for a restriction. Bis(pentabromophenyl)ether. ECHA, Helsinki

ECJ (2008) Joined cases C-14/06 and C-295/06

EEA (2001) Late lessons from early warnings: the precautionary principle 1896–2000. Environmental Issue Report No 22. European Environment Agency, Copenhagen

EEA (2013) Late lessons from early warnings: science, precaution, innovation. EEA Report No 1/2013. European Environment Agency, Copenhagen

EEC (1967) Council Directive 67/548/EEC on the classification, packaging and labelling of dangerous substances. OJ 196:1–98

EEC (1976a) Directive 76/769/EEC on restrictions of certain dangerous substances and preparations. OJ L 262:201–203

EEC (1976b) Council Directive 76/464/EEC on pollution caused by certain dangerous substances discharged into the aquatic environment. OJ L 129:23–29

EEC (1985) Council Directive 85/467/EEC amending for the sixth time (PCBs/PCTs) Directive 76/769/EEC on restrictions of certain dangerous substances and preparations. OJ L 269:17–19

EEC (1993) Council Regulation 793/93/EEC on the evaluation and control of the risks of existing substances. OJ L 84:1–75

Eriksson P, Jakobsson PE, Fredriksson A (2001) Brominated flame retardants: a novel class of developmental neurotoxicants in our environment? Environ Health Perspect 109:902–908
Eriksson P, Viberg PH, Jakobsson E, Orn U, Fredriksson A (2002) A brominated flame retardant, 2,2′,4,4′,5-pentabromodiphenyl ether: uptake, retention, and induction of neurobehavioral alterations in mice during a critical phase of neonatal brain development. Toxicol Sci 67:98–108
Eriksson J, Karlsson M, Reuter M (2010a) Technocracy, politicization, and non-involvement: politics of expertise in the European regulation of chemicals. Rev Policy Res 27:167–185
Eriksson J, Karlsson M, Reuter M (2010b) Scientific committees and EU policy: the case of SCHER. In: Eriksson et al (eds) Regulating chemical risks: European and global challenges. Springer, Dordrecht
Eskenazi B, Chevrier J, Rauch SA, Kogut K, Harley KG, Johnson C, Trujillo C, Sjödin A, Bradman A (2013) In utero and childhood polybrominated diphenyl ether (PBDE) exposures and neurodevelopment in the CHAMACOS study. Environ Health Perspect 121:257–262
EU (2010a) Commission Decision 2010/477/EU on criteria and methodological standards on good environmental status of marine waters. OJ L 232:14–24
EU (2010b) Commission Regulation (EU) 757/2010 amending Regulation (EC) No 850/2004 of the European Parliament and of the Council on persistent organic pollutants as regards Annexes I and II. OJ L 223:29–36
EU (2013) Directive 2013/39/EU of the European Parliament and of the Council amending Directives 2000/60/EC and 2008/105/EC as regards priority substances in the field of water policy. OJ L 226:1–17
European Commission (1994) Regulation 1488/94/EC laying down the principles for the assessment of risks to man and the environment of existing substances in accordance with Council Regulation 793/93/EEC. OJ L 161:3–11
European Commission (2001) Communication from the Commission to the Council, the European Parliament and the Economic and Social Committee. Community strategy for dioxins, furans and polychlorinated biphenyls. (COM(2001) 593 FINAL
European Commission (2005) Proposal for a Directive of the European Parliament and of the Council relating to restrictions on the marketing and use of perfluorooctane sulfonates (amendment of Council Directive 76/769/EEC). COM(2005) 618
European Commission (2009) European Union Strategy for the Baltic Sea Region. Communication from the Commission to the European Parliament, the Council, the European Economic and Social Committee and the Committee of the Regions. COM(2009) 248 FINAL
European Commission (2012) Report from the Commission to the European Parliament and the Council on the implementation of the Water Framework Directive (2000/60/EC). Commission Staff Working Document (2/2). SWD(2012) 379 FINAL
European Parliament (2006) Report on the proposal for a directive of the European Parliament and of the Council relating to restrictions on the marketing and use of perfluorooctane sulfonates. (COM(2005)0618 – C6-0418/2005–2005/0244(COD)). A6-0251/2006
Gerecke AC, Hartmann PC, Heeb NV, Kohler HPE, Giger W, Schmidt P, Zennegg M, Kohler M (2005) Anaerobic degradation of decabromodiphenyl ether. Environ Sci Technol 39:1078–1083
Giesy JP, Kannan K (2002) Perfluorochemical surfactants in the environment. Environ Sci Technol 36:146A–152A
Gilbert N (2011) Data gaps threaten chemicals safety law. Nature 475:150–151
Greaves AK, Letcher RJ (2013) Linear and branched perfluorooctane sulfonate (PFOS) isomer patterns differ among several tissues and blood of polar bears. Chemosphere 93:574–580
Helander B (1983) Reproduction of the white-tailed sea eagle *Haliaeetus albicilla* (L.). In: Sweden, in relation to food and residue levels of organochlorine and mercury compounds in the eggs. Academic thesis, Stockholm, Gotab
HELCOM (1982) Recommendation (3/1) regarding the limitation of the use of PCB's
HELCOM (1985) Recommendation (6/1) regarding the elimination of the use of PCBs and PCTs
HELCOM (1988) Declaration on the protection of the environment of the Baltic Sea. Helsinki, 15 February 1988

HELCOM (1998) Recommendation 19/5. HELCOM objective with regard to hazardous substances
HELCOM (2004) Recommendation 25/1 (Supersedes HELCOM Recommendation 6/1). Elimination of PCBs and PCTs
HELCOM (2007) Baltic Sea Action Plan. Adopted at HELCOM Ministerial Meeting, Krakow, Poland 15 November 2007
HELCOM (2010a) Hazardous substances in the Baltic Sea. An integrated thematic assessment of hazardous substances in the Baltic Sea. Proceeding 120B
HELCOM (2010b) HELCOM Ministerial Declaration on the implementation of the HELCOM Baltic Sea Action Plan. Moscow 20 May 2010
HELCOM (2010c). Recommendation 31E/1 (Supersedes HELCOM Recommendation 19/5). Implementing HELCOM's objective for hazardous substances
HELCOM (2013) HELCOM monitoring and assessment strategy
HELCOM, OSPAR (2003) Statement on the ecosystem approach to the management of human activities. First joint ministerial meeting of the Helsinki and OSPAR Commissions, Bremen, 25–26 June 2003
Helsinki Convention (1974) Convention on the protection of the marine environment of the Baltic Sea area
Helsinki Convention (1992) Convention on the protection of the marine environment of the Baltic Sea area
Herbstman JB, Sjödin A, Kurzon M, Lederman SA, Jones RS, Rauh V, Needham LL, Tang D, Niedzwiecki M, Wang RY, Perera F (2010) Prenatal exposure to PBDEs and neurodevelopment. Environ Health Perspect 118:712–719
Holmström KE, Järnberg U, Bignert A (2005) Temporal trends of PFOS and PFOA in guillemot eggs from the Baltic Sea, 1968–2003. Environ Sci Technol 39:80–84
Ilonka A, Meerts TM, van Zanden JJ, Luijks EAC, van Leeuwen-Bol I, Marsh G, Jakobsson E, Bergman Å (2000) Potent competitive interactions of some brominated flame retardants and related compounds with human transthyretin in vitro. Toxicol Sci 56:95–104
Jensen S (1966) Report of a new chemical hazard. New Sci 32:612
Jensen GK (2014) Health costs in the European Union. How much is related to EDCs? HEAL, Brussels
Joas M, Jahn D, Kern K (2008) Governance in the Baltic Sea region: balancing states, cities and people. In: Joas M, Jahn D, Kern K (eds) Governing a common sea. Environmental policies in the Baltic Sea region. Earthscan, London
Johansson N, Fredriksson P, Eriksson P (2008) Neonatal exposure to perfluorooctane sulfonate (PFOS) and perfluorooctanoic acid (PFOA) causes neurobehavioural defects in adult mice. Neuro Toxicol 29:160–169
Karlsson M (2010) The precautionary principle in EU and U.S. chemicals policy: a comparison of industrial chemicals legislation. In: Eriksson J, Gilek M, Rudén C (eds) Regulating chemical risks: European and global challenges. Springer, Dordrecht
Karlsson M, Gilek M, Udovyk O (2011) Governance of complex socio-environmental risks: the case of hazardous chemicals in the Baltic Sea. AMBIO 40(2):144–157
KEMI (Swedish Chemicals Agency) (2013a) Economic cost of fractures caused by dietary cadmium exposure. Report 4/13. KEMI, Stockholm
KEMI (Swedish Chemicals Agency) (2013b) Brandskum som möjlig förorenare av dricksvattentäkter (in Swedish). PM 5/13. KEMI, Stockholm
Kern K, Löffelsend T (2008) Governance beyond the nation states: transnationalization and Europeanization of the Baltic Sea region. In: Joas M, Jahn D, Kern K (eds) Governing a common sea. Environmental policies in the Baltic Sea region. Earthscan, London
Kooiman J (ed) (2003) Governing as governance. Sage, London
Kortenkamp A, Backhaus T, Faust M (2009) State of the art report on mixture toxicity. Final report of a project on mixture toxicology and ecotoxicology commissioned by the European Commission, DG Environment

Kratzer J, Ahrens L, Roos A, Bäcklin BM, Ebinghaus R (2011) Temporal trends of polyfluoroalkyl compounds (PFCs) in liver tissue of grey seals (*Halichoerus grypus*) from the Baltic Sea, 1974–2008. Chemosphere 84:1592–1600

Lau C, Anitole K, Hodes C, Lai D, Pfahles-Hutchens A, Seed J (2007) Perfluoroalkyl acids: a review of monitoring and toxicological findings. Toxicol Sci 99:366–394

Law RJ, Allchin CR, de Boer J, Covaci A, Herzke D, Lepom P, Morris S, Tronczynski J, de Wit CA (2006) Levels and trends of brominated flame retardants in the European environment. Chemosphere 64:187–208

Letcher RL, Bustnes JO, Dietz R, Jensse BM, Jorgensen EH, Sonne C, Verreault J, Vijayan MM, Gabrielsen GW (2010) Exposure and effects assessment of persistent organohalogen contaminants in arctic wildlife and fish. Sci Total Environ 408:2995–3043

Lindberg P, Sellström U, Häggberg L, de Wit CA (2004) Higher brominated diphenyl ethers and hexabromocyclododecane found in eggs of peregrine falcons (*Falco peregrinus*) breeding in Sweden. Environ Sci Technol 38:93–96

Lindstrom AB, Strynar MJ, Libelo EL (2011) Polyfluorinated compounds: past, present, and future. Environ Sci Technol 45:7954–7961

Linduska JP (1953) DDT. Det vilda måste skyddas mot insektsgifterna (in Swedish). Sveriges Natur 6. Svenska Naturskyddsföreningen, Stockholm

Looser R, Froescheis O, Cailliet GM, Jarman WM, Ballschmiter K (2000) The deep-sea as a final global sink for semi-volatile persistent organic pollutants? Part II. Chemosphere 40:661–670

Lopez-Espinosa MJ, Fletcher T, Armstrong B, Genser B, Dhatariya K, Mondal D, Ducatman A, Leonardi G (2011) Association of perfluorooctanoic acid (PFOA) and perfluorooctane sulfonate (PFOS) with age of puberty among children living near a chemical plant. Environ Sci Technol 45:8160–8166

Lopez-Espinosa MJ, Mondal D, Armstrong B, Bloom MS, Fletcher T (2012) Thyroid function and perfluoroalkyl acids in children living near a chemical plant. Environ Health Perspect 120:1035–1041

Lyons G (1999) Chemical trespass: a toxic legacy. WWF-UK, London

Magnusson K, Norén K (2012) The sensitivity of the Baltic Sea ecosystem to hazardous compounds. PM 9/12. Swedish Chemicals Agency, Stockholm

MMB (Miljömålsberedningen) (2012) Minska riskerna med farliga ämnen. Strategi för Sveriges arbete för en giftfri miljö (in Swedish). Delbetänkande av MMB. SOU 2012:38. Fritzes, Stockholm

Murawski (AM) (2007) Ten myths concerning ecosystem approaches to marine resources management. Mar Policy 31:681–690

NCM (Nordic Council of Ministers) (2004) Cost of late action – the case of PCB. TemaNord 2004:556. NCM, Copenhagen

NCM (Nordic Council of Ministers) (2014) The cost of inaction. A socioeconomic analysis of costs linked to effects on endocrine disrupting substances on male reproductive health. Tema Nord. Nordic Council of Ministers, Copenhagen, p 557

Nyberg E, Bignert A, Danielsson S, Mannio J, the HELCOM CORESET expert group for hazardous substances (2013a) Polybrominated diphenyl ethers (PBDE). HELCOM, Helsinki

Nyberg E, Bignert A, Danielsson S, the HELCOM CORESET expert group for hazardous substances (2013b) Perfluorooctane sulphonate (PFOS). HELCOM, Helsinki

Österblom H, Gårdmark A, Bergström L, Müller-Karulis B, Folke C, Lindegren M, Casisni M, Olsson P et al (2010) Making the ecosystem approach operational. Can regime shifts in ecological and governance systems facilitate the transitions? Mar Policy 34:1290–1299

Pierre J, Peters BG (2005) Governing complex societies – trajectories and scenarios. Palgrave Macmillan, Basingstoke

Pistocchi A, Loos R (2009) A map of European emissions and concentrations of PFOS and PFOA. Environ Sci Technol 43:9237–9244

Pyhälä M, Brusendorff AC, Paulomäki H, Ehlers P, Kohonen T (2007) The precautionary principle and the Helsinki Commission. In: de Sadeleer N (ed) Implementing the precautionary principle. Earthscan, London

Renner R (2001) Growing concern over perfluorinated chemicals. Environ Sci Technol 35:154A–160A

Renner R (2008) PFOS phaseout pays off. Environ Sci Technol 42:4618

Roos A, Greyerz E, Olsson M, Sandegren F (2001) The otter (*Lutra lutra*) in Sweden – population trends in relation to ΣDDt and total PCB concentrations during 1968–99. Environ Pollut 111:457–469

RPA (2004) Perfluorooctance Sulphonate. Risk reduction strategy and analysis of advantages and drawbacks. Final report. DEFRA, Norfolk

Rudén C, Hansson SO (2010) Registration, evaluation, and authorization of chemicals (REACH) is but the first step – how far will it take us? Six further steps to improve the European chemicals legislation. Environ Health Perspect 118:6–10

Rylander L, Strömberg U, Hagmar L (2000) Lowered birth weight among infants born to women with a high intake of fish contaminated with persistent organochlorine compounds. Chemosphere 20:1255–1262

Santillo D, Johnston P, Langston WJ (2001) Tributyltin (TBT) antifoulants: a tale of ships, snails and imposex. In: Harremoes P, Gee D, MacGarvin M, Stirling A, Keys J, Wynne B, Vaz G (eds) Late lessons from early warnings: the precautionary principle 1896–2000. Environmental issue report No 22. European Environmental Agency, Copenhagen, pp 135–148

Schantz SL, Gasior DM, Polverejan E, McCaffrey RJ, Sweeney AM, Humphrey HEB, Gardiner JC (2001) Impairments of memory and learning in older adults exposed to polychlorinated biphenyls via consumption of Great Lakes fish. Environ Health Perspect 109:605–11

SCHER (2005) Opinion on 'update of the risk assessment of bis(pentabromophenyl) ether (decabromodiphenylether)', Final environmental draft of May 2004. SCHER, European Commission, 18 March 2005

Scheringer M, Trier X, Cousins IT, de Voogt P, Fletcher T, Wang Z, Webster TF (2014) Helsingør statement on poly- and perfluorinated alkyl substances (PFASs). Chemosphere 114:337–339

Schuetze A, Heberer T, Effkemann S, Juergensen S (2010) Occurrence and assessment of perfluorinated chemicals in wild fish from northern Germany. Chemosphere 78:647–652

Selin H, VanDeveer SD (2004) Baltic Sea hazardous substances management: results and challenges. AMBIO 33:153–160

SNFA (2011) Redovisning av regeringsuppdrag rörande gränsvärden för långlivade miljöföroreningar i fisk från Östersjöområdet (in Swedish). Dnr 115/2010. Swedish National Food Administration (SNFA), Uppsala

Stapleton HM, Alaee M, Letcher RJ, Baker JE (2004) Debromination of flame retardant decabromodiphenyl ether by juvenile carp (*Cyprinus carpio*) following dietary exposure. Environ Sci Technol 38:112–119

Szlinder-Richert J, Barska I, Usydus Z, Grabic R (2010) Polybrominated diphenyl ethers (PBDEs) in selected fish species from the southern Baltic Sea. Chemosphere 78:695–700

TFEU (2007) Consolidated versions of the Treaty on European Union and the Treaty on the Functioning of the European Union. OJ C 326:1–390

Udovyk O, Rabilloud L, Gilek M, Karlsson M (2010) Hazardous substances: a case study of environmental risk governance in the Baltic Sea region. RISKGOV report to BONUS EEIG Programme. Södertörn University, Sweden

UN (1992) United Nations convention on biological diversity

UNEP (United Nations Environment Programme) (2006) Risk profile on perfluorooctane sulphonate. Report of the Persistent Organic Pollutants Review Committee on the work of its second meeting. UNEP/POPS/POPRC.2/17/Add.5. Geneva, 6–10 Nov 2006

UNEP (United Nations Environment Programme) (2013) Costs of inaction on the sound management of chemicals

Ward MH, Colt JS, Deziel NC, Whitehead TP, Reynolds P, Gunier RB, Nishioka M, Dahl GV, Rappaport SM, Buffler PA, Metayer C (2014) Residential levels of polybrominated diphenyl ethers and risk of childhood acute lymphoblastic leukemia in California. Environ Health Perspect 122:1100–1116

Young OR (1994) International governance. Protecting the environment in a stateless society. Cornell University Press, New York

Chapter 6
Oil Spills from Shipping: A Case Study of the Governance of Accidental Hazards and Intentional Pollution in the Baltic Sea

Björn Hassler

Abstract Despite most tankers being more technically safe than in the past, the increasing volume of transportation probably outweighs most, if not all, technical safety gains. Two major types of threats to the Baltic Sea environment caused by oil pollution are discussed in this chapter: accidental and intentional spills. It is shown that individual countries or coalitions have influenced governance outcomes in both areas. The introduction of double hull regulations by IMO was speeded up significantly by unilateral action taken by the USA and the EU. The move towards differentiated port controls has probably increased efficiency since it has made it possible to target substandard vessels. The Paris MoU has been important in ensuring coherent inspection practices. Intentional oil spills typically result from unlawful cleaning of tanks and engine rooms at sea. Flight surveillance and the No-Special-Fee system have been adopted to reduce oil spills. However, both mechanisms suffer from weaknesses caused by differences in countries' capacities and priorities. Flight surveillance intensity differs significantly among HELCOM member states, which makes it possible for tankers to avoid detection. The No-Special-Fee system has been only partially effective, due to varying interests and capacities of individual Baltic Sea countries, port authorities and ports.

Keywords Oil transportation • Double hulls • Oil spill • Port State Control • IMO

6.1 Introduction

Oil transportation has increased significantly in the Baltic Sea over the last couple of decades. There are 17 major oil ports in the Baltic Sea, and the volume of transported oil now exceeds 250 million tonnes yearly. It is expected that these amounts

B. Hassler (✉)
School of Natural Sciences, Technology and Environmental Studies, Södertörn University, 14189 Huddinge, Sweden
e-mail: bjorn.hassler@sh.se

M. Gilek et al. (eds.), *Environmental Governance of the Baltic Sea*, MARE Publication Series 10, DOI 10.1007/978-3-319-27006-7_6

will continue to grow. In parallel with the increased transportation of oil, vessels have on average become larger which has important implications in terms of worst-case scenarios of accidents.

The large amounts of oil being transported over the ecologically sensitive Baltic Sea create substantial environmental hazards. These risks are of two quite different and distinct types. On the one hand, there are accidental spills and on the other intentional spills caused by, for example, not taking proper care of polluted spill water. Accidents caused by collisions, fire, groundings, technical malfunctions, human error or other factors may result in large-scale oil spills. Depending on the amount of oil being carried, geographical area, type of oil, water temperature, icy conditions, wind and currents, damages to ecological systems, local tourism, recreation, real estate values and fisheries vary but could in worst-case scenarios be devastating. Fortunately, the safety of modern tankers in terms of its potential environmental costs has increased quite markedly during the last few decades due to increased emphasis placed on vessel construction and on-board installations such as advanced navigation equipment (Knudsen and Hassler 2011). In contrast, intentional spills are typically small in size, but because of their large number, the cumulated impact on the ecological integrity of the Baltic Sea is probably substantial, although detailed assessments of such spills are not available. Despite the illegality of these spills, they have been difficult to substantially curb. Due to ineffective monitoring in certain areas of the Baltic Sea and the difficulties to spot polluters and make them pay fines means that the practice of, for example, cleaning oily tanks at sea continues. However, as shown in this chapter, other ways to reduce intentional oil spills have been attempted with some success.

Contemporary marine governance in the area of shipping comprises a mix of carefully crafted hierarchical structures and horizontal interactions among a multitude of stakeholders (Bennet 2000; Knudsen and Hassler 2011; Mason 2003; Mitchell 1994). Compared with many other environmental risk areas, shipping is governed by a relatively clearly defined chain of command in terms of governance where all key global conventions have been placed under the umbrella of the UN organisations' IMO (International Maritime Organization) and ILO (International Labour Organization), after having been agreed upon by the member states. Ultimately, the implementation of adopted conventions, however, must be done at the national level, primarily in the form of Flag and Port State Control. It has been shown that although the Flag State according to UNCLOS has the formal responsibility for all ships flying its flag, Port State Control has become the most important mechanism to improve safety in global shipping (DeSombre 2006; Knapp and Frances 2008).

To understand actual governance outcomes, incentives as well as structures have to be carefully considered. Management of oil spill risks in the Baltic Sea is clearly affected not only by global conventions but often as importantly by the actions of the EU, by intergovernmental organisations such as HELCOM (Helsinki Commission) and by individual governments with strong interests and high capacity to take action on particular issues. Much of the focus in recent governance literature has been placed on interaction between institutions and stakeholders at different scales and the importance of including non-state stakeholders (Bach and Flinders 2004; Joas et al. 2008; Oberthür and Gehring 2006). In this chapter, it is argued that while global structures are important, contemporary multilevel and multi-actor governance patterns cannot be grasped without also including the roles played by individual countries and intergovernmental organisations. These actors are often driven by issue-specific interests but also bound by restrictions in terms of resource shortages and level of organisational skills. The focus of this chapter is on different measures that have been taken to reduce risks of accidental oil spills and strengthen operators' incentives to follow rules and norms on intentional pollution. It will be argued that the EU plays an important role both as regulator and enforcer in marine environmental governance, namely, in between the regional (HELCOM) and the global level (IMO/ILO). HELCOM, on the other hand, plays an important mediating role between national interests among the Baltic Sea countries and regulations in the EU and IMO. Furthermore, it will be argued that individual countries play important roles in specific governance situations, especially when they perceive strong national interests to be at stake. Mechanisms to facilitate proactive governments taking action when it is in their interest to do so could in certain situations contribute to improved and more sustainable governance.

The remaining part of this chapter is structured as follows. After a brief section on ecological and economic impacts of oil spills, we have a section on measures to reduce accidental spills and mitigate intentional spills. Examples of both command and control as well as incentive-based measures are given. The chapter concludes with a discussion on possible ways ahead, with a particular focus on what roles the EU and HELCOM can play in regional governance of marine oil transportations.

6.2 Ecological and Economic Consequences of Marine Oil Spills

The direct effects of large-scale oil spills are easy to observe. Seabirds and mammals especially may suffer and die in large quantities. Beaches may become unusable for recreation activities, and local fisheries can be devastated for a number of years. Laboratory studies have shown that oil can have deadly as well as sublethal effects upon organisms. Field studies after accidents have shown significant negative effects on affected ecosystems (National Research Council 2003). However, it is almost impossible to predict the long-term consequences from a significant oil spill, because of at least three different types of uncertainty.

First, the location of the spill is crucial (French-McCay 2009). Typically, the farther away from shores the spill occurs, the more time is available for limiting the geographical distribution of the spill, to fence off sensitive areas and set up cooperative cleanup schemes among authorities from different countries. This normally results in a more limited impact. Moreover, depending on the local ecological sensitivity and economic importance of the affected area, consequences may vary considerably.

Second, depending on the type of oil being spilled, weather conditions and sea currents, the ecological and social impacts can vary dramatically. For example, icy conditions make cleanup activities especially cumbersome. Third, the long-term effects of oil spills on complex ecosystems are not well known (National Research Council 2003). Whether or not oil sinks to the bottom and continues to affect specific communities and ecosystems for a long time depends on the type of oil and local conditions. Possibly, oil spills can increase physiological stress, reduce food supply and cause reproduction disturbances.

Taken together, these uncertainties make it impossible to in detail predict impacts of large-scale oil spills. Although political agreements may be reached on the importance of reducing risks, questions related to which particular measures are most cost-efficient cannot be definitely given as long as it is not possible to assess risks in detail.

The ecological impact of the many almost continuous small-scale intentional spills that take place in the Baltic Sea is quite different from the large-scale ones and more difficult to delineate. Although it could be expected that operators choose to clean tanks in ways that minimise risks of being spotted, existing data on where oil spills have been detected show a rather uniform distribution during the last 10 years along the major sea lanes (possibly with some clusters at the entrance of the Gulf of Finland and in the Great Belts) (HELCOM 2014). There are four kinds of major environmental concerns linked to intentional spills. First, many small spills occur close to ports as a result of loading and unloading and may cause disturbances to local ecosystems. Most of these spills are not intentional, but rather the result of improper procedures or human error. Despite this, they are typically classified as "intentional" in order to distinguish them from large-scale accidental spills.[1] Second, even though these smaller spills occur along the major shipping lanes in general, there seem to be some clusters where spills are more common. Environmental hazards typically increase the denser these clusters are. Third, where the spills take place is of importance. Comparably large spills in favourable weather conditions and far away from ecologically sensitive areas may not cause much observable harm, whereas small spills could cause substantial damage if they occur close to sensitive areas, in unfavourable weather conditions and in critical seasons when different marine-living species might be reproducing. Finally, the Baltic Sea is an ecologically sensitive sea because of its brackish water, slow water exchange with the North Sea and the halocline (salinity gradient) that reduces vertical mixing of water and thereby leads to reduced oxygen concentrations in deep basins.

[1] It could be argued, for example, that operational spills might be a better label than intentional spills, as the former places the focus on oil spills that result from everyday procedures rather than from accidents. However, since intentional spills is the term used in most of the literature, we have chosen to stick to this term when referring to small-scale spills caused by negligence, improper procedures or something similar.

6.3 Marine Environmental Safety Drivers

Probably the most important measure to reduce risks of large-scale accidental oil spills has been the requirement of double hulls. The double hulls of modern vessels means that the risk of oil spills are significantly reduced in case of a grounding or collision. In a 1992 amendment of MARPOL, it was stipulated that no large tankers (5,000 Deadweight tonnage or more) without double hulls could be ordered after 1993 unless IMO had recognised alternative designs that were deemed to be acceptable from an environmental safety perspective. However, since the conversion of single-hull tankers is complicated and the expected lifetime of vessels constructed before 1993 is up to 30 years, a complicated and prolonged phase-out process was elaborated by IMO. As will be described below, this phase-out process was subsequently speeded up, partly due to some large-scale accidents and partly due to unilateral action on prohibition of single-hull vessels by the USA and later by the EU.

A quite different type of environmental hazard is caused by operators' cutting corners. In order to save time and money, operators often get tanks illegally cleaned and flush machine rooms at sea, rather than in port where oily wastes could be taken care of properly. In some cases, these types of spills might not be intentional in a strict sense but nevertheless are caused by negligence or improper procedures. The number of observed illegal spills has been decreasing over the last two decades, although the number of unrecorded cases is probably substantial. While the number of flight surveillance hours to combat illegal pollution increased during the 1990s and then levelled off during the 2000s, the number of detected spills declined over the last 10 years from approximately 500 in the late 1980s to around 130 in the last couple of years (HELCOM 2014). The introduction of satellite monitoring by EMSA (European Maritime Safety Agency) over the last 5 years has been instrumental in directing aerial and marine surveillance. The long-term ecological effects of this diffuse form of pollution of the Baltic Sea are not known in detail but could potentially be substantial.

To understand why some measures to reduce oil spills in the Baltic Sea have been more successful than others, why some measures are better handled at regional levels and others through global conventions and why some countries seem to be more proactive than others, it is important to consider differences in interests and capabilities (Hassler 2010). First, environmental hazards that are global in nature typically require global conventions, as regulations at regional and subregional levels tend to invite free riding and market distortions where actors act strategically to avoid costs and reap benefits without contributing to the realisation of collective goods (Keohane 2002). For example, as we discuss further below, tanker construction needs to be regulated globally. Although it could be suggested that the Flag State responsibility ought to be enacted more forcefully as a way to counteract collective action dilemmas, the emergence of open registries has made this very difficult, if not impossible. On the other hand, measures taken by individual countries to improve environmental safety that primarily give local effects may escape the "Tragedy of the commons" (Hardin 1968). It has been shown that measures such as improved hydrographical measurements and navigation charts, not only in domestic waters but also in collaboration with neighbouring countries, have been successfully undertaken (Hassler 2011). The global features of marine oil transportation are not what is important, but rather the fact that the country carrying the costs is also the prime beneficiary of the undertaken project. Whether to make this investment or not thus becomes an issue of domestic cost-benefit analysis which does not need to turn into a tragedy of the commons. Basically, it then becomes a question for the authorities in that country to decide how to finance this investment.

Second, different countries will benefit disproportionally from most kinds of pollution control. Countries with extended coastlines close to major marine traffic lanes, for example, are typically more vulnerable to oil spills. It could be expected that these countries would be more proactive vis-à-vis stricter regulations, everything else being equal. Not only would these countries be expected to undertake measures having primarily local effects as described above, but proactive positions in relation to international regulation would also be anticipated. Considering that lobbying for stricter international regulation is typically not costly in relation to what could be gained in case of successful interventions, there is no reason to expect barriers for collective action in these cases, given that domestic interests are sufficiently strong. This is most likely one important reason why Sweden and Finland have been strong proponents of stricter international regulations with regard to environmental hazards in the Baltic Sea, not only when it comes to marine oil transportation but to most other threats to the integrity of Baltic Sea ecosystems as well. In a similar manner, but conversely, countries with large stakes in marine oil transportation could be more hesitant to accept costly measures to reduce risks of oil spills. This would, according to the logic of collective action rationality, especially apply to stricter international regulations since these might threaten economic interests. On the other hand, local improvement in, for example, port facilities or hydrographical charts could be more appealing for countries in geographically vulnerable situations.

Third, and finally, not only interests matter, but capability as well. Countries with more resources, know-how and experience could be expected to be more proactive in relation to risk prevention as well as the build-up of impact reduction and cleaning up capability compared with countries that have less resources and experience with these issues. This aspect has considerable relevance in the Baltic Sea region as there is a marked difference in resource availability and administrative experience between the former Soviet Union states on the one hand and the Nordic states and Germany on the other. This has had significant implications not only for national environmental management, transposition of EU regulations and implementation of international agreements but also in a wider, regional sense. It has been shown, especially during the first decade of independence in the 1990s, that the Nordic states influenced the Baltic States by giving targeted support to strengthen their administrative capability and assisting them in their preparation for becoming EU members (Hassler 2003a).

It is clear in summary that perceiving environmental governance of marine transportation as globalised and therefore uniformly prone to collective action problems is overly simplistic. To understand what would be the main drivers in effective marine governance, it is necessary to disentangle individual actors' – governments', sector organisations', operators', NGOs' and others' – interests in specific cases and analyse how these interests contribute to collective outcomes. In some cases, notably where costs and benefits from an undertaking fall upon a single actor, problems related to collective action should not be expected.[2]

6.4 Undertaken Measures to Reduce Oil Spills in the Baltic Sea

Turning now to the type of measures that have been taken to reduce risks related to marine oil transportation, it is clear that large-scale accidental oil spills and intentional limited spills are similar in the sense that management suffers from significant amounts of uncertainty. However, different types of uncertainty in these areas mean that different types of measures need to be taken to reduce oil pollution in the Baltic Sea. For example, regarding accidental spills vessel construction is of crucial importance. Since most vessels are sailing in waters in different parts of the world, uniform rules are typically required to induce compliance. When it comes to intentional spills, on the other hand, most spills are operational in nature, implying that monitoring and surveillance is required. In addition to the distinction between accidental and intentional spills, it is helpful to distinguish between *command and control* measures (binding regulations) and various *incentive schemes* aimed at altering actors' behaviour in more environmentally friendly directions (to reduce accidental risks and intentional spills). The latter are predominantly based on economic incentives (e.g. differentiated tariffs) but could also comprise other forms of incentives, such as benchmarking and environmental labelling. The distinction between command and control mechanisms and incentive schemes is important since it directly ties to the above discussion on actors' varying interests. Aspects that need uniform, global regulation typically call for global command and control mechanisms, mainly because of the difficulties of administrating a global incentive scheme, such as a uniform environmental tax on fuels. The main problem with command and control regulation is low levels of implementation and compliance.[3] Whereas the key rationale of command and control measures is uniform application, the opposite could be said to hold for incentive-based schemes. Although the

[2] It could be argued that problems related to collective action may re-emerge at the national level, since various actors at the domestic level can be assumed primarily to promote particular rather than joint interests. However, since the hierarchical structure is much stronger at the national level compared to the international, collective action aspects are in most cases not as serious at the national level.

[3] That some members refrain from being members of IMO, which is the most important global authority on marine affairs, is not a major problem as the number of members now has reached 170 (IMO 2011a). While not all member countries have signed and ratified all individual conventions, the even bigger problem is the lack of implementation and compliance of ratified conventions.

Table 6.1 Matrix showing four examples of different categories of marine oil spill control

	Command and control	Incentive schemes
Accidental spills	Double hulls	Differentiated Port State control
Intentional spills	Flight surveillance	Integrated port reception fees

framework as such should be applied uniformly, the bearing idea is to differentiate among actors depending on individual behaviour. Preferred behaviour should be awarded, while unwanted choices should be penalised. To make such a scheme effective, it is imperative to know the main drivers of key actors.

In Table 6.1 the distinctions are shown between, on the one hand, accidental and intentional spills and, on the other hand, command and control mechanisms and incentive schemes. In the rest of this section, examples of each of these four combinations are discussed. It should be noted that (a) the cases given are far from exhaustive but should be viewed as illuminating examples and (b) some examples may exhibit both command and control and incentive-based mechanisms, something that will be discussed below.

6.5 Accidental Spills

6.5.1 Command and Control Measures: EU (and US) Influence over the Phasing out of Single-Hull Tankers

In the early 1600s, Hugo Grotius formulated the idea of a *Mare Liberum*, the freedom of the seas, a principle stating that all states have the right to use the sea for transportation and trade. This idea was later codified in UNCLOS (United Nations Convention of the Law of the Seas). The bearing idea in UNCLOS is that innocent passage should always be allowed (Article 17) and that it is the responsibility of the Flag State to make sure that vessels carrying its flag follow all valid international agreements (Article 94). It is moreover the Flag State that is responsible for carrying out investigations when incidents occur.

IMO (the International Maritime Authority) is the main intergovernmental authority governing the seas. IMO was established in 1948, but it was not until 1958 that the first convention entered into force and the work could begin. During the first decades of its operation, IMO's focus on environmental protection stressed the need for regulating and monitoring of intentional spills and operating procedures (Mitchell et al. 1999). Rules were devised with regard to maximum oil contents in spill water and how far from the coast such pollution was allowed (Hassler 2010). However, it soon became apparent that it was not possible to effectively control operational procedures at sea, mainly because of the vast geographical areas that had to be covered but also because of the inadequacy of Flag State control (Knudsen and Hassler 2011). When it became clear that monitoring and Flag State enforcement were not effective, more focus was put on easily controllable requirements.

The most obvious of such requirements were directly related to vessel construction and retrofitting. When new vessels are ordered, the purchaser needs to make sure that it complies with the most recent IMO conventions; otherwise the Classification Society will not grant the needed permit. In other words, there is a rather effective and efficient mechanism for making sure that new vessels comply with existing regulations. Interestingly, private actors – Classification Societies – play a crucial role in making these command-and-control measures effective.

The institutionalised system of Port State Control is undertaken on selected vessels so as to ensure that required safety installations are operational. The major weakness of this system is that there are no guarantees that installed safety equipment actually are used in accordance with proper procedures. A major determining factor of whether such equipment is used or not is the extent the operator has economic or other incentives not to use installed equipment. When proper use is not costly, or even beneficial to the operators, it could be expected that intended procedures are adhered to. On the other hand, when operators gain from cutting corners by not using installed equipment, technical requirements are typically not sufficient.

The ongoing international phasing out of single-hull tankers is probably the single most important initiative that has been taken to increase environmental safety in relation to accidental large-scale oil spills. In what follows, the importance of phasing out single-hull tankers will be elaborated upon. Attention is also given to how individual countries may take unilateral action in order to protect what is perceived to be of national interest and how large-scale accidents can create momentum for adoption of stricter regulation.

The first initiative to phase out single-hull tankers was taken unilaterally in 1990 by the USA (Oil Pollution Act; OPA 90) as a direct consequence of the 1989 *Exxon Valdez* accident. The US ban meant that neither new nor old tankers with single hulls would be allowed to call on US ports after 2005. IMO reacted to the US ban in 1992 when it accepted an amendment of MARPOL that stated that large tankers (over 5,000 Deadweight tonnes) must have double hulls if ordered after July 1993 (MARPOL, Annex I, Regulation 19). However, the issue of how to phase out single-hull tankers which were in use without creating too much disruption in marine transportation was more difficult to agree upon. Initially, it was decided in IMO that existing tankers should either be converted or taken out of service before they were 30 years old (MARPOL, Annex I, Regulation 20).[4] EU authorities, faced by a situation where all single-hull tankers would be denied access to US ports in 2005 following the serious Erika (1999) and Prestige (2002) oil spills, realised that there was a high probability that the unilateral decision by the USA would result in a redirection of single-hull ships from the USA to other parts of the world (Summaries of EU

[4] The 30 year limit was, however, not absolute but could be adapted to, e.g. bottlenecks in shipyards' capability to handle conversions to double hulls and to whether the vessel had segregated ballast tanks or not. Owners could therefore ask IMO for extension periods for their vessels on individual basis. However, because of the difficulties in retrofitting existing vessels, it was assumed that older tankers would rather be taken out of service than converted.

Legislation 29-08-2011; Aksu et al. 2004). EU ports could thus be expected to receive a larger share of single-hull tankers than before (Regulation (EC) No 417/2002). As a reaction to this threat, the EU enacted a unilateral speeding up of the phasing-out timetable that had already been established in the US OPA 90, stipulating that no Category 1 tankers would be allowed to enter any port in any EU country or carry the flag of any EU member country after 2005 (Regulation (EC) No 417/2002). Vessels constructed in 1980 or earlier would not be allowed after 2003 and those constructed in 1981 not after 2004. Category 2 and 3 tankers were given a deadline of 2010, with the newest vessels being given deadlines furthest into the future. However, vessels in Category 2 and Category 3 that were older than 15 years in 2005 were subjected to enhanced surveys – so-called Condition Assessment Scheme (CAS) – especially targeting structural weaknesses in single-hull vessels (Regulation (EC) No 417/2002). Faced by considerable pressure from the EU (Höfer 2003), IMO attempted a speed-up of the single-hull phase-out and a revised schedule entered into force in 2003 (MARPOL Regulation 13 G). In December 2003 additional revisions were made, stipulating that Category 1 tankers (large vessels not having segregated ballast tanks) had to be phased out no later than 2005. Category 2 tankers (large tankers with segregated ballast tanks) and Category 3 tankers (small vessels) were required to have double hulls from 2010 onwards, not 2015 as previously decided (IMO 2012-06-29). The Flag State could, however, still give permission for Category 2 and 3 tankers according to what was stipulated in the IMO CAS (Condition Assessment Scheme), to continue operation until 2015 or until they were 25 years old. This brief review of the phasing out of single-hull vessels is an interesting example of how dominant players (the USA and EU) interact with an intergovernmental authority (IMO) on command and control measures related to environmental safety.[5] The USA and EU pushed forward the phasing out of single-hull oil tankers, thus facilitating a global phasing out through IMO mechanisms. They were able to do this due to their large share of the world market. They had to do this because of domestic political pressures to increase safety in marine oil transportation and perceived dangers in being negatively affected by unilateral action by others.[6]

[5] However, it should be noted that although UNCLOS regulations on the right of Port State Control have been important in the regime for in-phasing of double hull vessels, double hulls are probably not a panacea for improved safety. It has been shown that these constructions often are harder to inspect and that inadequate technical solutions and poor maintenance may result in only limited safety enhancements.

[6] Although relevant IMO regulations do not have complete global coverage, MARPOL Annex I/II has been ratified by 150 countries (2011b), representing more than 99 % of global merchant shipping tonnage (IMO 2011a).

6.5.2 Incentive Schemes: The Case of Selective Port Inspections in the Baltic Sea and Other European Waters

Environmental governance schemes based on incentive mechanisms rather than on command and control regulations typically use taxes or subsidies to alter behaviour in preferred directions. These schemes have become widely used especially at the national level mainly because they increase economic efficiency when designed appropriately, that is, when externalities are internalised into companies' and organisations' budgets. Theoretically, although often difficult to achieve in real situations where lack of information, uncertainty and strategic behaviour interfere with management objectives, taxes could be tuned to perfectly offset externalities such as negative environmental effects from industrial production or transportations. This would make it rational for targeted actors to reduce pollution until the marginal cost of environmental side effects equals marginal pollution reduction costs.

Unfortunately, preconditions for successfully establishing economic incentive schemes at the international level are radically different. It has proven difficult to find robust systems to tax use of ecosystem services and natural resources in the international commons as they are not under any single country's jurisdiction. The highly globalised marine transportation sector is no exception and is indicative of the fact that almost no environmental taxes have been successfully and uniformly applied in this area. However, there are other than economic ways to influence actors' behaviour through altered incentives. Similar to taxes and subsidies, these other incentive-based mechanisms ideally should be constructed so that behavioural change among targeted actors result in as large positive environmental effects as possible in relation to costs inferred. In other words, actors that behave well should come out better than those behaving not so well.

The modern Port State Control is an interesting example of such an incentive-based mechanism where regional Memoranda of Understandings (MoUs) have been instrumental in coordinating port inspections. In this brief example, the focus is placed on the Paris MoU, an organisation established in 1982 and comprising today 26 European member countries plus Canada and the EEC.[7] Until quite recently, about 25 % of visiting ships were randomly selected for inspection, but in January 1, 2011, the so-called NIR (New Inspection Regime) was implemented (Paris MoU 2012c). In the NIR regime, all vessels are assessed when calling at Paris MoU ports and those that are believed to be more likely to have safety deficiencies selected for inspection. In order to build up a legitimate basis for the selection of risky vessels, a centralised database has been established under the auspices of EMSA (European Maritime Safety Authority). The ship risk profile is updated daily, and specific vessels are selected based on parameters such

[7] Apart from the Paris MoU, there are nine other regional sister authorities throughout the world: Abuja MoU, Black Sea MoU, Caribbean MoU, Indian Ocean MoU, Mediterranean MoU, Riyadh MoU, Tokyo MoU, Asia Pacific Region and Viña del Mar Agreement Latin American Region.

as previous history of detected deficiencies, age and company performance.[8] All vessels that have been classified as high-risk ships have to make a port call at least 72 h before planned arrival in order to facilitate expanded inspections by the port authorities. Failure to report notifications as is required for all visiting ships may result in selection for inspection, irrespective of prior ship profile. It may also result in fines depending on national regulations in the Port State. The ports where these ships are heading will then carry out an inspection according to criteria stipulated in IMO regulations (Paris MoU 2012a). If the inspectors find major deficiencies, the vessel may be detained until these have been corrected. Repeated detentions may in turn result in banning from all Paris MoU member countries' ports for a specified period of time.

Inspections take time and therefore create direct costs for companies. Being classified as a high-risk ship may moreover result in indirect costs for the companies as this may impact upon insurance costs as well as the willingness of cargo owners to do business with these vessel owners. The switch from the earlier procedure of random controls to targeted controls of vessels has thus resulted in strengthened incentives for operators to achieve a low-risk profile. The only way vessel owners can achieve this is to improve their record by reducing the probability of being found to have deficiencies. According to preliminary data from Paris MoU, more detentions took place, despite fewer inspections, in 2011 as compared to previous years, which seems to indicate a higher degree of inspection effectiveness (Paris MoU 2012b).

The Paris MoU has in different ways given the port inspection mechanism increased potency. This has to a significant extent been possible due to the capacities of Member States and their ability to collaborate. This form of regional cooperation is of crucial importance in international governance since formal enforcement mechanisms are lacking. One way to give these incentive-based mechanisms more clout is to make sure that the information on ship deficiencies and detentions not only are made public but are also made as easily accessible and widespread as possible. Paris MoU is primarily doing this through its homepage (http://www.parismou.org), where statistics as well as individual vessel records, risk profiles and other data are published. Finally, statistics on Flag State and Classification Society performance are published as well. Depending on the number of inspections and detentions, Flag States and Classification Societies are classified in white/grey/black lists easily accessible to all stakeholders.

[8] The company performance indicator measures the frequency of found deficiencies among a company's complete fleet of vessels and may result in rankings from very low to high performance.

6.6 Intentional Spills

6.6.1 Command and Control: Aerial Surveillance and Monitoring of Baltic Sea Waters

Whereas IMO has moved from vessel monitoring at sea towards focusing on vessel construction and maintenance, the situation is slightly different with regard to regional management of intentional oil spills in the Baltic Sea. All relevant IMO regulations are also valid in the Baltic Sea region, and many of them moreover have been incorporated into the Helsinki Convention. Due to the smallish size of the Baltic Sea, serious attempts have been made to monitor ship movements in order to detect oil spills. There are mainly three reasons why this form of monitoring could be important at regional and subregional levels but less effective in the large open seas of the world. First, the number of countries that need to cooperate on monitoring schemes is limited which facilitates joint action. In the case of the Baltic Sea region, the longstanding record of HELCOM being a key regional intergovernmental organisation facilitates reaching of agreements (Hassler 2003b). Second, the fact that most of the major Baltic Sea shipping lanes are within individual countries' exclusive economic zones (EEZs) and territorial waters means that these countries may find themselves to be in vulnerable positions in case of oil spills. This creates incentives for individual countries to deter operators from polluting nearby but still in international waters (Hassler 2008). Third, and finally, the limited size of the Baltic Sea makes aerial surveillance in collaboration with coastguard patrols effective enough to be worthwhile. Since April 2007, monitoring has been given more muscle through the EMSA CleanSeaNet satellite surveillance. However, even though satellite surveillance may seem like a powerful tool, HELCOM data from 2012 shows that from the 185 satellite detections of suspected spills, only 13 were later confirmed as mineral oil spills, while more than half (93) were never checked at all (HELCOM 2012).

The main responsibility of aerial surveillance rests with individual countries. Agreements have thus been reached within HELCOM that air surveillance should be undertaken by all member countries and that statistics of the number of flight hours, detected oil spills, confirmed oil spills and similar kinds of data should be reported to HELCOM. HELCOM thereafter compiles reports and makes these available to member countries as well as to the general public. In addition to this, intensified, joint surveillance efforts are undertaken once or twice a year under the programme Coordinated Extended Pollution Control Operation (CEPCO). During CEPCO operations, a 24-h surveillance scheme is carried out. On some occasions so-called Super-CEPCO operations have also been undertaken that last for 6–10 days in an attempt to improve data on pollution and reduce the number of spills. The details of these operations remain classified until after the surveillance has been carried out in order to avoid strategic actions by operators.

It is clear that ambition levels among different HELCOM member countries have varied considerably despite the fact that HELCOM Recommendation 12/8 adopted as far back as 1991 spoke about joint responsibility to collaborate and contribute to effective aerial surveillance. As can be seen in Fig. 6.1, the number of yearly flight hours varies considerably among HELCOM member countries, although there has been a slight upward trend over time at the aggregate level. For example, Sweden in almost all years has had more flight hours recorded than all the other countries together. Russia, in contrast, has only recorded ten flight hours between 1993 and 2011. Even though it is reasonable to expect countries with long coastlines and major sea routes to have more flight hours than others, the overall impression is that compliance with flight hour commitments is uneven.

Despite the uneven implementation of the flight surveillance schemes, available data seem to indicate that the number of intentional spills has declined over the last decade, in spite of the significant increases in marine traffic at the Baltic Sea (Fig. 6.2).

Although the interpretation of HELCOM data may seem straightforward – improved command and control measures through satellite and aerial and surface monitoring have led to decreased levels of intentional oil spills – uncertainties remain regarding the number of spills as well as what drivers have been most important in the alleged reduction of intentional spills. It may very well be the case that increased surveillance has led to fewer intentional oil spills in the last decade. Most likely this is the main explanation for the declining trend shown in Fig. 6.2. However, the number of observed spills is most likely considerably lower than actual spills. It is reasonable to assume that vessel operators choose a particular time (during

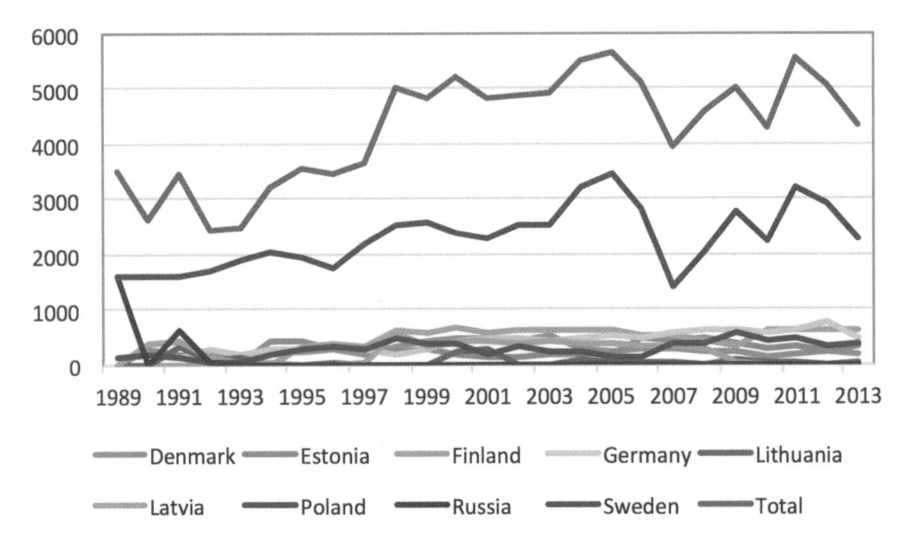

Fig. 6.1 Number of yearly surveillance flight hours between 1989 and 2013 (Adapted from HELCOM 2014) (Scores of zero flight hours in this figure represent no reported flight hours for that year or that zero flight hours have been reported)

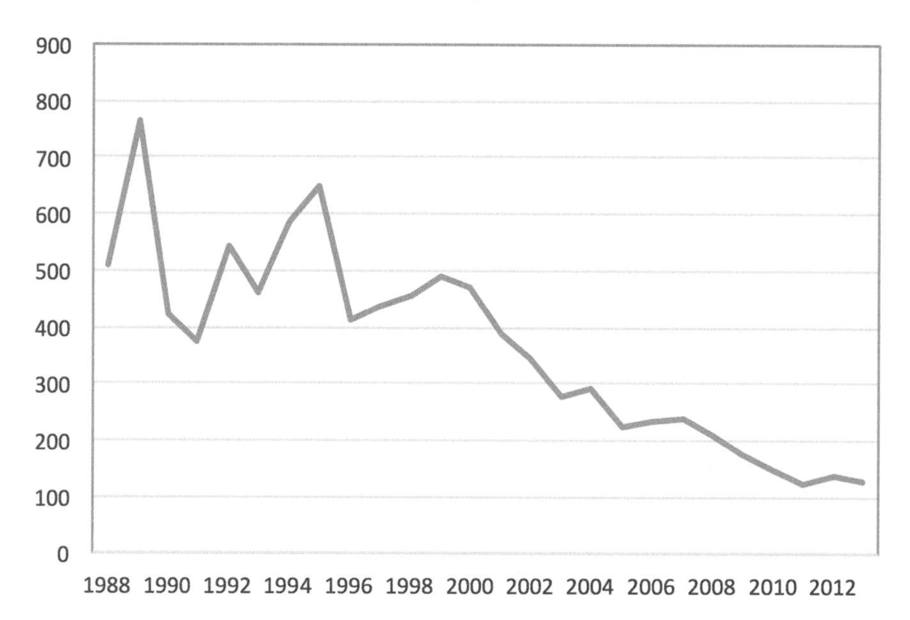

Fig. 6.2 Number of recorded intentional oil spills between 1988 and 2013 (Adapted from HELCOM 2014)

darkness) and place (areas where surveillance is known to be lax) to make intentional spills. If this is correct, the number of undetected oil spills could be substantial. Possible partial solutions to this problem could be more effective satellite surveillance, increasing the number of flight hours in darkness and using and intensifying the use of joint surveillance efforts that are not announced beforehand (CEPCO operations). Moreover, if we are in fact witnessing a reduction in the number of spills, it is likely that these command and control measures are not the only reasons for this reduction. At least two additional factors are probably important. First, as tanker fleets in the Baltic Sea are modernised, the pollution from, for example, flushing of machine rooms could be expected to decrease as spill water can be better taken care of in modern tankers. Moreover, when vessels are modernised and better adapted to existing reception facilities in the ports, the incentives to clean tanks at sea are weakened, as less time than before is needed for cleaning tanks in port. Second, the introduction of more effective reception facilities and the No-Special-Fee system (elaborated on below) means that no additional charges are made for cleaning tanks in port. Incentives for cleaning them at sea are thus reduced.

Two concerns remain regarding the effectiveness of aerial surveillance. First, member countries are responsible for covering their respective air space. This means that the engagement shown most likely will continue to vary in accordance with national capacity and priorities. Second, detection of oil spills typically do not lead to identification of the perpetrator. Only in 12 % of the detected spills has the polluting ship been identified (HELCOM 2014). Moreover, even when the polluter is identified, it is far from certain that this leads to a conviction, and if it does, penalties are typically low. Despite the use of HELCOM AIS – a land-based tracking system that makes it possible to track all large ships in real time – and the Seatrack Web oil drift forecasting system (STW) that facilitates assessment of propagation patterns of oil spills, identification effectiveness of polluters remains poor. This obviously weakens incentives for operators to cut corners by, for example, cleaning tanks at sea.

6.6.2 *Incentive Schemes: The Baltic Sea No-Special-Fee System*

One way to reduce temptations for operators to intentionally pollute by, for example, cleaning tanks at sea is to reduce costs for abiding by MARPOL conventions that require depositing oily wastes in dedicated port reception facilities. Despite the fact that port reception facilities have been on the agenda of all MEPC meeting for the last 35 years since the advent of MEPC 3 (Mikelis 2010), facilities and operational procedures are still far from adequate in the ports of many MARPOL member countries.[9]

From the perspective of IMO, "MARPOL provisions require the government of each party to ensure the provision of adequate port reception facilities without causing undue delay" (IMO 1999). Moreover, according to MEPC, "The ability of ships to comply with the discharge requirements of MARPOL depends largely upon the availability of adequate port reception facilities…" (MEPC 2007). However, what constitutes "adequate port reception facilities" is almost impossible to define since the kind of facilities necessary depends on what types of vessels regularly call in ports. In smaller ports simple barrels might be sufficient, whereas in many of the larger ports, facilities have to be constructed that cannot only harbour large amounts of wastes but also treat different oil residues in different ways. The issue of how to delimit "…without causing undue delay" is not a question that is easy to solve. It is clear that every extra hour a large tanker has to stay in port costs the operator a significant amount of money. Therefore, temptations to clean tanks at sea will prevail as long as it is quicker and easier than doing so in a port, and the risk of being caught while polluting is negligible.

In order to put pressure on Port States and stimulate investments in port reception facilities, IMO has underlined the responsibility of Flag States to ensure that the IMO report format is distributed to all vessels carrying their flag. This format should be used by the master of the vessel to notify the Flag State as well as IMO and the relevant Port State when reception facilities are not appropriate or when there have been undue delays. Port States are then expected to "… ensure the provision of

[9] MEPC – the IMO Marine Environment Protection Committee.

proper arrangements to consider and respond appropriately and effectively to reports of inadequacies, informing IMO and the reporting flag State of the outcome of their investigation" (MEPC 2007).

EU's strategy in this area has been similar to the one on Port State Control referred to above. EU Directive 2000/59/EC reiterates the regulation pertaining to Port State facilities stipulated in MARPOL.[10] However, this directive is different from the MARPOL Convention in two ways. First, it is only applicable to EU member countries. However, despite being applicable to EU Member States only, this does not mean that vessels from other countries are exempted. On the contrary, all vessels (except for war ships and non-commercial ships owned or operated by a State) have to abide by the procedures and regulations on handling of waste that are stipulated in the directive when entering a port of an EU country (Article 3). Second, by turning the MARPOL regulation into an EU directive, enforcement mechanisms are substantially strengthened. Port States that do not follow the directive could be taken to court.

It is stipulated in Directive 2000/59/EC that each port in every Member State has to establish a waste reception and handling plan (Article 5). This plan in turn has to be approved by the government of the Member State, and every third year at least, it has to be reapproved by the government. The Member State is moreover responsible for monitoring the implementation of the Waste reception and handling plan. In order for port authorities to prepare for inspections or other procedures, operators are required to notify the port they are calling on, where and when residues were left in their previous port visit and how much waste is still on-board. The operators are moreover required to leave residues in the port before leaving, unless they can show that they can adequately store the waste on-board. Failing to do so could mean that the ship is not allowed to leave the port (Article 7). Finally, the port is required to ensure that it covers the costs of its reception facilities. In order to do this, all visiting vessels are required to pay a certain part of the reception costs, irrespective of whether they use the port's facility or not.[11] The Commission has quantified this amount to be at least 30 % of total reception costs (Directive 2000/59/EC). Apart from this, fees are set in accordance with the amount of waste delivered. However, "...fees may be reduced if the ship's environmental management, design, equipment and operation are such that the Master of the ship can demonstrate that it produces reduced quantities of ship-generated waste" (Article 8c).

Against this background of initiatives taken to improve port reception facilities at the global and EU level in order to reduce the temptation to pollute at sea, the regional HELCOM No-Special-Fee system is interesting (HELCOM

[10] Directive 2000/59/EC entered into force on December 28, 2000. The deadline for countries to implement the directive was December 28, 2002.

[11] Interestingly, this may be somewhat at odds with one of the most important principles in modern environmental protection – the Polluter Pays Principle (PPP) – since operators that do not need to use port reception facilities (because of, e.g. installed on-board equipment to manage operation spill) still have to pay for this service.

Recommendation 19/8).[12] The HELCOM No-Special-Fee system builds on MARPOL requirements on waste reception facilities in ports and the EU demand on cost coverage. However, it focuses more on actual incentives facing operators. This system is defined as:

> ...a charging system where the cost of reception, handling and disposal of ship-generated wastes, originating from the normal operation of the ship... is included in the harbour fee or otherwise charged to the ship irrespective of whether wastes are delivered or not" (HELCOM Recommendation 28E/10, paragraph 1.1).

In other words, this system implies that all ships have to pay for reception, handling and disposal of residues, even if they have no residues to account for. The bearing idea behind the No-Special-Fee system is that the port fee should not be related to the amount of residues the vessel leaves in port. This would mean that – at least in theory and not taking the extra time in port needed for proper waste management into consideration – the operator would not gain from, for example, cleaning tanks at sea, since the handling of the residues is free (i.e. included in the port fee). This does not necessarily mean that all ships have to pay the same fee. A more reasonable approach is that ships pay according to a selected parameter that could be expected to vary with the average amount of waste, but not with the amount of waste deposited by a specific ship. However, it is not stipulated precisely in HELCOM Recommendation 28E/10 what particular parameter should be used. Gross tonnage as available from the ship's Data Sheet would be an easy measure, but type of cargo, number of staff and the quality of on-board installations for waste management as stated in the Recommendation could also be used. Independent of what measures are used, the principles of fee calculation should be "fair, transparent and non-discriminatory to all ships" and ensure a high degree of legitimacy. Moreover, the collected fees should exclusively be used for costs related to waste reception in the port (HELCOM Recommendation 28E/10, paragraph 4). Finally, in order to avoid market distortions, it has been stated that "The Contracting States involved shall make the necessary efforts in order to implement a harmonised fee system simultaneously in the ports of the Baltic Sea as well as in the North Sea Regions" (HELCOM Recommendation 28E/10, paragraph 5).

Despite the fact that governments are urged to periodically submit reports on the implementation of the No-Special-Fee system, few recent authoritative assessments of the system's effectiveness have been published. According to a joint report from HELCOM member countries on the implementation of the Helsinki Convention of 2005, only three (Denmark, Finland and Germany) out of seven audited countries

[12] HELCOM Recommendation 19/8 on waste reception in ports was later superseded by new recommendations where additional types of wastes and garbage were included in the No-Special-Fee system. The latest, valid recommendation on these issues is now (August 2012) "HELCOM Recommendation 28E/10: Application of the No-Special-Fee System to Ship-Generated Wastes and Marine Litter Caught in Fishing Nets in the Baltic Sea Area" which was adopted in November 15, 2007.

had fully implemented the No-Special-Fee system.[13] According to more recent sources, there are indications that the full implementation of this system still seems to be problematic (Jensen 2011). One suggested reason for this is that some ports, because of the competition among them to attract business, have chosen to charge those using reception facilities more than others. Another factor based on observations is that ports have charged extra fees when pumping sludge outside of regular office hours (Jensen 2011). It is quite clear that the No-Special-Fee system works better in ports where more effective reception facilities have been installed and governments and responsible authorities have both the will and capacity to improve compliance.

6.7 Discussion

Contemporary marine governance is often depicted as comprising multilevel and multi-actor interactions, competing knowledge claims and evolving patterns of co-management, where stakeholders and users play increasingly important roles in overall governance. However, while this is true with regard to fisheries, eutrophication and protection of biodiversity, Baltic Sea shipping is characterised by hierarchical governing structures where IMO acts as a global regulatory hub, EU as a stakeholder and enforcer, HELCOM as an important interface between individual governments and intergovernmental organisations at regional and global scales and governments as key stakeholders typically promoting issue-specific national interests. Despite its globalised nature, modern shipping is to a considerable extent governed by intergovernmental organisations practising a policy mix comprising command and control measures as well as mechanisms to reduce gaps between operators' economic incentive structures and politically agreed upon regulations.

It has been illustrated in four brief examples that both the EU and HELCOM have played important roles in Baltic Sea marine governance. However, their roles have been markedly different from each other. An interesting role that the EU has played at times is that of an enforcer of global conventions at the EU level. The EU has used its legal regulatory instruments and directives, for example, to enforce a faster phasing out of single-hull vessels within the Union and thereby also speed up the global IMO phasing-out process. Another example was when the EU turned the recommended 25 % port inspection frequency by IMO into a mandatory requirement in all EU ports. The EU has, in other words, been able to strengthen global conventions internally and sometimes even influence global governance.

In contrast, HELCOM has no other governance mechanisms at its disposal besides for the Helsinki Convention and the Recommendations which both are built upon consensual decision-making and do not allow for legal enforcement. Despite this, the example of the No-Special-Fee system shows that it can be possible to take

[13] Estonia, Latvia, Lithuania and Poland had partially implemented the No-Special-Fee system, whereas Sweden and Russia were not audited.

regulations at higher levels (IMO and EU requirements on port reception facilities) one step further. This could in turn serve as a testing ground and inspiration for similar initiatives at higher levels (e.g. EU). In fact, as has been discussed, a No-Special-Fee system can be implemented throughout the EU building on the example provided by HELCOM. Another aspect of regional governance is that an organisation such as HELCOM can sometimes facilitate subregional collaboration. We illustrated above that joint aerial surveillance for oil pollution was carried out under HELCOM's CEPCO initiative. Although it was also shown that the number of flight hours varies considerably and that not all countries are equally interested in regional cooperation on this matter, it is clear that subregional cooperation among a limited number of reasonably like-minded countries could be valuable and possible for a smaller group. It is typically not the case that participation of all on equal terms is necessary. Sometimes it could be more efficient to let a group of proactive countries take the lead, possibly putting some pressure on others to follow suit.

It is often said that political decision-making and regulatory structures always should be placed at the "appropriate level", that is, at the level that corresponds to the scale of the problem at hand. This is true, but easier said than done. The real challenge is to identify the appropriate level not so much for the sector as such (i.e. shipping) but more importantly for the detailed aspect that needs to be regulated. Moreover, finding the appropriate scale also means that varying opportunities are opened up for different regulatory mechanisms that define how effective and efficient they could be. Although the phasing out of single-hull tankers needs global conventions, unilateral action by dominant powers such as the USA and EU could speed up things when global regulation drags on for too long.

In contrast to command and control mechanisms involved in the phasing out of single-hull tankers, changing incentives have been instrumental to the selective Port State Control and the No-Special-Fee system. The key here has been to look closer at the incentives facing the main actors and trying to change these incentives to induce behaviour that results in safer shipping. It should be noted that in all these cases, there is a global convention at the bottom formalising what has been possible to reach consensus on. Regulations at lower levels or using other means always have to be undertaken within the boundaries set by these higher level agreements in order to avoid inconsistencies. But there are often different ways to improve the enforcement of these regulations and adapt them to regional conditions without threatening the consistency of the global convention, as shown by especially the Port State Control and the No-Special-Fee system examples.

Finally, addressing incentives and capability in marine governance is not only important in relation to operators but also in relation to governments as well. It is no coincidence that Sweden has so many more flight hours than Russia. Sweden and Finland especially have been proactive in implementing regional environmental regulation in the Baltic Sea. Likewise, it is not surprising that the Baltic States have had more problems than the Scandinavian countries in the implementation of the No-Special-Fee system. It is quite natural that proactive countries such as Sweden and Finland tend to exaggerate the amount of common interest in addressing threats to the Baltic Sea environment. The simple reason for this is that it might be in their strategic interest to do so in order to induce less interested countries to do more. However, officially recognising differences in national interests and capability might be important in order to set the stage for negotiations under more realistic preconditions. This could mean that proactive countries will need to contribute more than others to monitoring, surveillance, training in Port State Control procedures and part financing of port reception facilities in countries less rich in resources and other initiatives. This might be preferable to negotiation breakdown or severe implementation gaps. Here, HELCOM as well as EU arenas could prove instrumental in facilitating the finding of workable agreements among the Baltic Sea governments.

Acknowledgements This chapter is based on research undertaken within the research programme "Environmental Risk Governance of the Baltic Sea" (2009–2015, Michael Gilek programme coordinator). The research involved research teams from Södertörn University in Sweden, Åbo Akademi University in Finland, Dialogik/Stuttgart University in Germany and Gdansk University in Poland. The funding came from the Foundation for Baltic and East European Studies and the European Community's Seventh Framework Programme (2007–2013) under grant agreement no. 217246 made with the joint Baltic Sea research and development programme BONUS, as well as from the German Federal Ministry of Education and Research (BMBF), the Swedish Environmental Protection Agency, the Swedish Research Council FORMAS, the Polish Ministry of Science and Higher Education and the Academy of Finland. I wish to express my warm appreciation to these institutions for enabling me to conduct this research, to all participants in the research programme that directly or indirectly provided useful input, to all informants for sharing their experiences and to the two reviewers for their constructive comments on an earlier version of the chapter.

References

Aksu S, Vassalos D, Tuzcu C, Mikelis N, Swift P (2004) A risk based design methodology for pollution prevention and control. RINA International Conference on Design and Operation of Double Hull Tankers, London

Bach I, Flinders M (eds) (2004) Multi-level governance. Oxford University Press, Oxford

Bennet P (2000) Environmental governance and private actors: enrolling insurers in international maritime regulation. Polit Geogr 19(7):875–899

DeSombre ER (2006) Flagging standards: globalization and environmental, safety, and labor regulations at sea. MIT Press, Cambridge

French-McCay DP (2009) Oil spill impact modeling – development and validation. Environ Toxicol Chem 23(10):2441–2456

Hardin G (1968) The tragedy of the commons. Science 162(3859):1243–1248

Hassler B (2003a) Science and politics of foreign aid – Swedish environmental support to the Baltic States. Kluwer Academic Publishers, Dordrecht

Hassler B (2003b) Protecting the Baltic Sea – the Helsinki convention and national interests. In: Stokke OS (ed) Yearbook of international co-operation on environment and development. Earthscan, London

Hassler B (2008) Individual incentives, collective action, and institutional cooperation – Sweden and the case of Baltic Sea oil transports. J Environ Policy Plan 10(4):339–358
Hassler B (2010) Global regimes, regional adaptation; environmental safety in Baltic Sea oil transportation. Marit Policy Manag 37(5):489–504
Hassler B (2011) Accidental versus operational oil spills from shipping in the Baltic Sea – risk governance and management strategies. AMBIO 40:170–178
HELCOM (2012) Annual 2012 HELCOM report on illegal discharges observed during aerial surveillance, 2013, Helsinki
HELCOM (2014) Annual 2013 HELCOM report on illegal discharges observed during aerial surveillance, 2013, Helsinki
Höfer T (2003) Tanker safety and coastal environment: Prestige, Erika and what else? Environ Sci Pollut Res 10(1):1–5
IMO (1999) Comprehensive manual on port reception facilities, 2nd edn. Ashford Press, London
IMO (2011a) Member states, IGOs and NGOs. Available from: http://www.imo.org/About/Membership/Pages/Default.aspx. Accessed 27 Jun 2012
IMO (2011b) Construction requirements for oil tankers. Available from: http://www.imo.org/OurWork/Environment/PollutionPrevention/OilPollution/Pages/constructionrequirements.aspx#1. Accessed 31 Jul 2012
IMO (2012) Tanker safety – preventing accidental pollution. Available from: http://www.imo.org/blast/mainframe.asp?topic_id=155#double. Accessed 29 Jun 2012
Jensen C (2011) Recommendations for improved and harmonized waste management on board and in ports. The Regional Council in Kalmar County, Kalmar, June 2011
Joas M, Detlef J, Kern K (eds) (2008) Governing a common sea: environmental policies in the Baltic Sea region. Earthscan, London
Keohane RO (2002) Power and governance in a partially globalized world. Routledge, London
Knapp S, Frances PH (2008) Econometric analysis to differentiate effects of various ship safety inspections. Mar Policy 32(4):653–662
Knudsen OF, Hassler B (2011) IMO legislation and its implementation: accident risk, vessel deficiencies and national administrative practices. Mar Policy 35(2):201–207
Mason M (2003) Civil liability for oil pollution damage: examining the evolving scope for environmental compensation in the international regime. Mar Policy 27(1):1–12
MEPC (2007) Revised consolidated format for reporting alleged inadequacies of port reception facilities. MEPC.1/Circ.469/Rev.1, 13 Jul 2007
Mikelis N (2010) IMO's action plan on tackling the inadequacy of port reception facilities. Powerpoint presentation presented at the workshop "Ships' waste – time for action" in Brussels, 14 October 2010 and organised by EUROSHORE and FEBEM-FEGE
Mitchell RB (1994) Intentional oil pollution at sea: environmental policy and treaty compliance. MIT Press, Cambridge, MA
Mitchell RD, McConnell ML, Roginko A, Barrett A (1999) International vessel-source oil pollution. In: Young O (ed) The effectiveness of international environmental regimes – causal connections and behavioral mechanisms. MIT Press, Cambridge, MA
National Research Council (2003) Oil in the sea III – inputs, fates and effects. The National Academic Press, Washington, DC
Oberthür S, Gehring T (eds) (2006) Institutional interaction in global environmental governance – synergy and conflict among international and EU policies. MIT Press, Cambridge, MA
Paris MoU (2012a) Ship risk profile. Available from: http://www.parismou.org/Inspection_efforts/Inspections/Ship_risk_profile/#503dec47-c60f-466c-9887-a2354722cf19. Accessed 2 Mar 2012
Paris MoU (2012b) A port state control preview. Available from: http://www.parismou.org/Organization/30_years_of_Paris_MoU_on_Port_State_Control/2012.07.02/A_Port_State_Control_preview.htm. Accessed 31 Jul 2012
Paris MoU (2012c) New inspection regime: port state control reporting obligations. Available from: https://www.parismou.org/Content/PublishedMedia/613aa272-374d-4ead-9ba1-6acf5dc5a25c/Reporting%20obligation.pdf. Accessed 31 Jul 2012
Summaries of EU Legislation (29-08-2011) Maritime safety: accelerated phacing-in of double-hall tankers. Available from: http://europa.eu/legislation_summaries/transport/waterborne_transport/l24231_en.htm. Accessed 07 May 2015

Part II
Cross-Case Analysis of Key Environmental Governance Challenges

Chapter 7
The Ecosystem Approach to Management in Baltic Sea Governance: Towards Increased Reflexivity?

Magnus Boström, Sam Grönholm, and Björn Hassler

Abstract This chapter analyses the governance structures linked to the marine environment of the Baltic Sea. The purpose is to assess whether current developments of the governance structures have a potential to take into account requirements of an Ecosystem Approach to Management (EAM). We use the concept of *reflexive governance* to understand key components and weaknesses in contemporary governance modes, as well as to elaborate on possible pathways towards a governance mode more aligned with EAM. The reflexive governance framework highlights three elements: (1) acknowledgement of uncertainty and ambiguity; (2) a holistic approach in terms of scales, sectors and actors; and (3) acknowledgement of path dependency and incremental policy-making. Our analysis is based on a comparative case study approach, including analysis of the governance in five environmental risk areas: chemical pollution, overfishing, eutrophication, invasive alien species and pollution from shipping. The chapter highlights an existing governance mode that is ill-equipped to deal with the complexity of environmental problems in a holistic manner, with systematic attention to uncertainty, plurality of values, ambiguity and limited knowledge, while also pointing at important recent cognitive and institutional developments that can favour pathways towards reflexive governance and consequently EAM.

Keywords Baltic Sea • Ecosystem approach to management • Holistic approach • Marine governance • Reflexive governance

M. Boström (✉)
School of Humanities, Education and Social Sciences, Örebro University, SE-701 82 Örebro, Sweden
e-mail: magnus.bostrom@oru.se

S. Grönholm
Åbo Akademi University, Turku, Finland
e-mail: sgronhol@abo.fi

B. Hassler
School of Natural Sciences, Technology and Environmental Studies, Södertörn University, 14189 Huddinge, Sweden
e-mail: bjorn.hassler@sh.se

M. Gilek et al. (eds.), *Environmental Governance of the Baltic Sea*, MARE Publication Series 10, DOI 10.1007/978-3-319-27006-7_7

7.1 Introduction

Our perceptions of what constitutes adequate marine environmental governance have undergone substantial changes during the last decades. From having been firmly based on biophysical problem framings, scientific risk assessments and end-of-pipe solutions, integrated ecosystem approaches to management, stakeholder participation and adaptive governance are now what is on the table.

Marine environmental governance of the Baltic Sea has furthermore become more multifaceted than before, involving not only governments, international government organisations (IGOs) and other public bodies but increasingly also transnational networks of public authorities at substate levels, sector organisations, industry and non-governmental organisations (NGOs) (Hassler et al. 2011). The inclusion of non-public actors has to a considerable extent co-evolved with increased use of economic instruments such as differentiated levies, taxes and subsidies aimed at changing actors' behaviour through altered economic incentives (Hassler et al. 2011). Moreover, several scholars regard various forms of interaction between international conventions and EU directives as important parts of contemporary environmental governance (Oberthür and Gehring 2006; Stokke 2001; Young 2002).

Adding to the complexity of environmental governance, there is an increasing understanding among scholars that risk, uncertainty and ambiguity are key factors of the problems at hand (Gilek et al. 2011; Hassler et al. 2013; Lidskog et al. 2009; Renn 2008; Voss et al. 2006a; Walker and Shove 2007). Because of the systemic nature of marine ecosystems, a holistic approach needs to complement, and probably sometimes replace, the traditional emphasis on single species and tightly circumscribed risk assessments that exclude all except formal experts from assessment and management (Hassler et al. 2011, 2013). In many cases it is more reasonable to talk about uncertainty than risk (as conventionally understood within risk assessment), as limited knowledge in combination with stochastic processes makes estimation of probabilities for different outcomes difficult or not even possible (Hassler et al. 2011). Moreover, the notion of ambiguity implies that there is no obvious normative standpoint about preferable actions that comes from a particular risk assessment. Different stakeholders may have different interpretations of the perceived risks based on varied ethical, religious and political frames of reference.[1]

This chapter analyses the governance structures of the marine environment of the Baltic Sea. The primary goal is to assess whether current developments of the governance structures have a potential to take into account requirements implied by an ecosystem approach to management (EAM). When we talk about governance structures, we refer to the regulatory frameworks and institutions governing the marine environment of the Baltic Sea. The origin and evolution of large changes in governance structures is of particular interest. These changes are often initiated by

[1] Ambiguity or ambivalence refers to an object or event that simultaneously can belong to two contrasting categories (e.g. simultaneously loving and hating someone). A risk, such as driving fast or eating tasty but unhealthy food, often has both positive and negative denotations.

governmental and other actors who have particular interests and capabilities with regard to specific environmental domains. Structural and discursive factors also shape conditions under which processes of change occur. We use the concept of *reflexive governance* as the main analytical vehicle to understand key components in contemporary governance modes, as well as to elaborate on possible pathways to make Baltic Sea environmental governance more sustainable.

This chapter is structured as follows: As EAM is increasingly seen as the manner in which governance should take place, we first provide a brief description of this management approach. We then give a brief description of the historical background to existing institutions and regulatory structures so as to understand contemporary Baltic Sea environmental governance structures and potential future pathways. Therefore, this chapter provides a section on key regional organisations, initiatives and regulations and highlights the increasing influence of the EU. This general background is then followed by an account of the concept of reflexive governance, which in turn forms the theoretical basis for the following extended analysis of contemporary governance structures with empirical examples from the areas of chemical pollution, overfishing, eutrophication, invasive alien species and pollution from shipping (see other chapters in the book). Finally, in Sect. 4, we discuss how the concept of reflexive governance can shed light on existing weaknesses of governance structures and how increased reflexivity could contribute to a more sustainable Baltic Sea.

In terms of method, our analysis is based on a comparative case study approach. The empirical data is drawn from five case studies conducted in the RISKGOV project. These case studies are documented in (Haathi et al. 2010; Hassler et al. 2010; Lemke et al. 2010; Sellke et al. 2010; Udovyk et al. 2010) and further analysed in Part I of this book (Hassler 2016; Karlsson and Gilek 2016; Karlsson et al. 2016; Sellke et al. 2016; Smolarz et al. 2016). The case studies were undertaken during 2009–2014 and were all based on secondary literature, analysis of key policy documents, semi-structured qualitative interviews with key actors in the respective risk areas (in total about 100) and participatory observations during conferences, workshops, consultations and round-table stakeholder conferences. A common analytical and methodological framework guided all five case studies which ensured comparability of the cases (see Gilek et al. 2016; Hassler et al. 2011 for further details).

7.2 The Ecosystem Approach to Management (EAM)

The Ecosystem Approach to Management (EAM), which endorses a multi-sector approach, has recently been embraced by key policy and regulatory actors. EAM underlies the Water Framework Directive (WFD), the Marine Strategy Framework Directive (MSFD), the EU Sixth Environmental Action Programme and even more so the EU's Baltic Sea Strategy and Maritime Policy.

EAM is a response to today's deepening biodiversity crises and promotes a broad holistic approach that goes beyond traditional management based on single species and single sectors (Curtin and Prellezo 2010; Grumbine 1994, 1997). EAM aims to provide a system of management that views the ecosystem as a whole, embracing ecological boundaries and integrity in such a way that all relevant ecosystem drivers and their impacts are considered in relation to their effects on ecosystem functioning and societal parameters. In broad terms EAM calls for adaptive, precautionary and knowledge-based measures across national and administrative borders to protect and restore key ecological functions of the environment (Backer et al. 2010). It is accordingly argued in EAM that management cannot be limited by administrative or political boundaries but needs to be delineated by appropriate biophysical boundaries. In this respect, institutional change and innovation is vital in dealing with and responding to institutional barriers (cf. Backer et al. 2010; Österblom et al. 2010). Emphasis is also given to interagency cooperation and promoting cooperation between various affiliated international, national and subnational management agencies. Furthermore, calls have been made for cross-sector integration, cross-sector resource management and integration of information across disciplines and harmonising institutions (Barnes and McFadden 2007; Berghöfer et al. 2008; Murawski 2007). While acknowledging interaction between ecology and society, and the important role of stakeholder inclusion, the 'social dimension' (including environmental justice, quality of life, social cohesion, work etc.) of sustainable development has so far garnered less attention than the environmental dimension in EAM thinking (see Dreyer et al. 2011).

EAM acknowledges the complexity of ecosystems and the uncertainty related to management, realising that all factors affecting ecosystems are not well understood and may never be. In EAM, scientific knowledge is perceived to be provisional, because management is viewed as a learning process, incorporating the results of previous actions and allowing management to adapt to uncertainty. Learning about ecosystem processes and the interpretation and responses to ecosystem feedback at multiple scales requires different types of knowledge, not only scientific biological knowledge but also other types of knowledge gained from experience with concrete ecosystems (farming, fishing, recreation etc.) (Barnes and McFadden 2007; Galaz et al. 2008). Broadening of stakeholder engagement is a key component in EAM, not only to formalise existing knowledge but also to complement poor governance and data so as to find consensual solutions and legitimise the management process (Arkema et al. 2006; Curtin and Prellezo 2010; Murawski 2007; Tallis et al. 2010).

The concept of reflexive governance, which will be introduced later, shares several features of this perspective. Indeed, the very notion of reflexivity, which has to do with learning based on self-reference and self-critique, has affinities with the kind of adaptive management that EAM calls for. However, we are not just adding the theory of reflexive governance because it appears to fit well with EAM. Rather, to understand the conditions (barriers and opportunities) for a governance system to adopt features of EAM, we argue that it is important to learn from social scientific understanding of governance structures and processes. We also have to take into account a theory that helps us understand institutional change and inertia in a realistic way. We consider the theory on reflexive governance relevant and useful for that purpose (see also Hassler et al. 2013).

7.3 Evolving Environmental Governance

Before the collapse of the Soviet Union, regional collaboration was more or less restricted to governmental discussions within the framework of the Helsinki Commission (HELCOM) and scientific collaboration at expert levels. The cold war placed strict limits on what was politically possible to achieve. According to some observers, HELCOM was primarily a means for the Soviet Union to find a relatively neutral ground for airing declarations mainly targeted at US audiences (Hjorth 1992). However, the early role of HELCOM as a first institutionalisation of regional collaboration for the protection of the Baltic Sea environment should not be underestimated (Hassler 2003b). During HELCOM's first two decades of existence, important scientific contacts were developed especially between researchers from Nordic countries and the Soviet Union. These Baltic Sea networks were important building blocks and resulted in several recommendations made by the still rudimentary, epistemic communities on concrete measure that could be taken with regard to ecological threats to the Baltic Sea environment.

Following the collapse of the Soviet Union, the declaration of independence of Baltic states and the democratisation of Poland in the early 1990s, four overarching changes affecting Baltic Sea environmental governance took place in the following two decades.

First, the liberation of the Baltic states and democratisation of Poland sparked off extensive bilateral support programmes from primarily Sweden, Finland and Denmark (Hassler 2003a). These programmes often involved governmental actors, such as municipalities and county boards, as well as non-governmental actors, for example, CECAMS (Centre for Environmental Science and Management) and Environmental Consultancy and Monitoring Centre, both from Latvia.[2] In order to establish collaborative schemes, twinning agreements were made between donor and recipient countries (Hassler et al. 2011). In these agreements, a public authority or private company in the donor country collaborated with a counterpart in the recipient country in order to stimulate transfer of knowledge and know-how (Hassler et al. 2011). Financing was typically secured through government funds or international financial institutions such as NEFCO (Nordic Environment Finance Corporation), EBRD (the European Bank for Reconstruction and Development) and IBRD (the World Bank). In addition to these collaborative schemes, transnational initiatives were taken, such as collaboration among political parties, labour unions, NGOs and sector organisations. For example, political parties in Sweden and Finland collaborated with sister organisations in the Baltic states, and labour unions similarly had projects targeting areas such as gender equality and workplace democracy. Typically, these undertakings were partly financed by public sources (Hassler 2003a). Taken together, these new forms of interaction all contributed to

[2] CECAMS and the Environmental Consultancy and Monitoring Centre were both later on transformed into Latvian public agencies (Hassler 2003b).

the strengthening of Baltic Sea governance and the possibility of deepened collaboration.

Second, the expansion of the EU, with Sweden and Finland becoming members in 1995 and the Baltic states and Poland in 2004, profoundly affected governance patterns as the Baltic Sea had now almost become a sea internal to the EU (Hassler et al. 2011). Only Russia among the Baltic Sea coastal states is not a member of the EU. National legislation had to be adapted to the common EU regulatory framework, and enforceable directives were added to the arsenal of steering mechanisms. Sizeable funds from EU's regional programmes were also a bonus. Apart from the direct effects the regulations and transposed directives have had on national legislation pertaining to marine environmental safety among the Baltic Sea EU member countries, new opportunities have evolved for strategically scaling up national and regional initiatives to the EU level in order to increase impact. Moreover, global conventions too have been scaled down, and provisions from these conventions added to the EU arsenal of enforceable law (Hassler et al. 2011). The vertical dynamics of Baltic Sea environmental governance has thus increased significantly over the last two decades.

EU programmes and EU-funded projects of various kinds moreover have exerted considerable and increasing influence on marine governance in the Baltic Sea region. The largest initiative so far has been the *Baltic Sea Region 2007–2013 Programme* comprising a total funding of 223 million euros plus national co-financing.[3] Apart from Baltic Sea EU member countries, Russia, Belarus and Norway are members of this Programme. Among its four priority areas, the third one is focused on the ecological integrity of the Baltic Sea in a wide sense (BSR programme 2012). The BSR programme is built on individual projects supported within the programme. Among the 80 projects funded, 20 are specifically targeting the Baltic Sea environment and concern areas such as maritime safety, climate change, eutrophication and aquaculture.[4] It is clear that the substantial interactions among project members and associated partners – researchers, public authorities, companies, sector organisations, NGOs and other stakeholders – significantly affect contemporary Baltic Sea regional governance. Collaborations from this programme will continue with a new Interreg BSR Programme for 2014–2020.[5]

Taken together, the combined effects from EU regulation on the environmental status of the Baltic Sea, the increased dynamics in terms of interacting regulatory schemes at different levels, the growth of direct contacts between various stakeholders and EU authorities and the formulation of EU macro-regional programmes

---

[3] While a co-financing level of 25 % is required for the more affluent Nordic countries and Germany, 15 % is required from the Baltic states and Poland, whereas only 5 % is the minimum in the case of Belarus (eu.baltic.net 2012).

[4] A project database comprising all BSR Programme projects is available at http://eu.baltic.net/Project_Database.5308.html?

[5] For more info, see http://eu.baltic.net/Future_period_2014_2020.26029.html. Accessed 15 December 2014.

have created a very different governance setup in the Baltic Sea region. The impact of this governance setup, however, has so far only been rudimentarily researched.

Third, the scope of what constitute environmental concerns has been broadened over the last few decades (Kern 2011). At first, they were narrowly perceived as management problems that in most cases could be dealt with by experts and through various technical solutions. Now what is asked for, not the least in several EU directives and policies, are multidimensional analyses, stakeholder involvement and system-wide management approaches. Stakeholder participation is typically emphasised not only as important in relation to public legitimacy from a normative standpoint but also as a means to bring in more knowledge to policy-making and make implementation more effective (Dreyer et al. 2011; Stirling 2009). Several EU directives have in this regard become increasingly important in the over-all governance of the Baltic Sea. A clear trend is that these directives are now more oriented towards holistic approaches to regulation and management. The Urban Wastewater Treatment Directive (91/271/EEC) and the Nitrates Directive (91/676/EEC) were both decided upon before the EU was enlarged in 1995. This meant that neither Sweden and Finland nor the Baltic states and Poland shaped the directives but nonetheless had to more or less adapt them when becoming members. The Water Framework Directive (2000/60/EC) has had profound influence on, for instance, new forms of collaboration, since it calls for river basin management domestically and regionally and includes key stakeholders in decision-making processes. REACH, the EU regulation on chemicals (EC/1907/2006), does not directly target pollution of the seas and has many gaps (numerous chemicals are unregulated). It is, however, of significant importance in relation to the impact of various chemicals on marine ecological systems as it requires producers and users to improve registration, evaluation and information provision regarding substances and preparations of high concern. The Marine Strategy Framework Directive (2008/56/EC), finally, is the most recent directive on marine environments. Although it is yet too early to tell what impact it eventually will have, as with the other recent directives, its major ambition is to achieve Good Environmental Status by 2020 in all EU marine waters. Compared to earlier attempts to protect marine environments, the MSFD comprises a more holistic approach with an emphasis on adaptive management and participatory deliberation and centre-state biodiversity conservation (Dreyer et al. 2014). A similar trend towards holistic and inclusive approaches in general and the Ecosystem Approach to Management in particular is apparent for regional strategies and action plans. For example, in the 2007 HELCOM Baltic Sea Action Plan (BSAP), references to the ecosystem approach, integrated management and stakeholder participation are made as well to the need to better understand interactions between social and ecological systems. All member countries are required to submit national implementation plans for BSAP, covering all major areas related to marine governance of the Baltic Sea, including land-based pollution sources. However, to what extent BSAP will be able to profoundly influence management practice down to the local level and within implementing authorities is too early to tell. So far, implementation in Sweden has largely been traditional in the sense of not including elaborate strategies on how to improve stakeholder participation, deliberative processes and adaptive governance (Hassler et al. 2013).

The three above-mentioned trends all concern changes in government-centred structures, programmes and approaches, which in turn introduce new conditions for environmental governance structures and processes in the region. Fourth, an ongoing transnationalisation from below can be observed, which sets forth new conditions for both environmental governance of the Baltic Sea and reflexive governance. Various actors from different countries have developed forms and capacity for much tighter interaction and collaboration. Such interaction is facilitated by umbrella organisations for municipalities (e.g. Union of the Baltic Cities), or environmental NGOs (e.g. Coalition Clean Baltic), or by already transnational environmental NGOs such as WWF that connect different offices and establish new ones.[6] By such increasing transnational collaboration and capacity building, business organisations, NGOs, municipalities and other non-state actors strengthen their abilities (viz. organisational, cognitive and other types) to participate effectively in environmental governance. In addition, they are able to increasingly by-pass national governments and approach EU authorities directly (sometimes through European-level umbrella organisations, such as the European Environmental Bureau), a policy track which they often find to be more efficient.

The four broad trends towards increased interaction among various stakeholders in the area of marine governance in the Baltic Sea region, the more institutionalised form and impact of EU governing and the evolving consensus on adopting systemic rather than particular approaches to environmental management add up to substantial changes in the preconditions for further marine governance improvements in line with EAM. In order to analyse these trends more thoroughly, we examine them from the perspective of reflexive governance.

7.4 Reflexive Governance

Reflexivity basically means self-reference. The concept of reflexive governance accordingly points to the need for strategic thinking on how to build institutions and governance modes that are, on the one hand, forward-looking and able to cope with uncertainties and other reasons for complex problems and, on the other hand, backward-looking and entailing self-critical scrutiny of the previous and current mode of governance (Voss et al. 2006a). Reflexivity entails reflecting upon the very governance process itself including its conditions: i.e. the extent to which existing boundaries, constitutions, discourses, policies, regulations, science-policy interactions and organisational arrangements produce and reproduce this generation of environmental problems. Reflexive governance thus includes a continuous self-oriented examination of positive and negative outcomes (substantive dimensions;

[6] See Boström et al. (2015), Kern and Löffelsend (2004) and Van Deever (2011). See also Jönsson et al. (2016).

sometimes understood as single-loop learning) as well as on how governance itself is related to these outcomes (structural/procedural dimension; sometimes understood as double-loop learning).[7]

We apply this theoretical framework to shed light on our five focused cases of marine environmental governance. We ask if relevant arrangements and discourses have been developed that enable such reflexivity. A number of factors may hinder or facilitate such reflexivity but we do not attempt an exhaustive analysis. Based on the existing work on reflexive governance, we develop an ideal type that consists of three elements that appear essential for reflexive governance and which guide our analysis of the Baltic Sea marine environmental governance arrangements. These elements are:

1. Acknowledgement of uncertainty and ambiguity
2. Holistic approach in terms of scales, sectors and actors
3. Path dependency and incremental policy-making

The first element is acknowledgement of uncertainty and ambiguity. Reflexive governance involves recognition that global and local sustainability problems are extremely complex, uncertain and ambiguous and need to be handled as such. Problem handling requires transgressing existing cognitive and institutional boundaries (Voss and Kemp 2006). Reflexive governance thus stresses that problems often cannot be 'solved', in the strict sense of the word, only handled. New problems or trade-offs tend to appear after decisions are made. Decisions tend to involve compromises. Surprises appear over and over again. The crucial question is if governance has any potential to continuously respond to such surprises.

The concept reflexive governance relates to Ulrich Beck's theory of the world risk society and reflexive modernisation (1992, 1994). Beck connects late modern risks to how societal spheres and institutions (technology, science, politics, the state, the economy) in the 'simple' or 'first' modernity operated according to the Enlightenment Project. Instrumental, rational problem-solving during 'simple' modernity relied on a cognitive and institutional approach in which uncertainty, complexity and ambiguity were disregarded. Problem-solving was specific and straightforward – based only on single-loop learning – and the goal was maximising the control of social and economic development. However, this approach inevitably leads to many unintended, negative consequences, at first called 'side effects'. As these are multiplied and increasingly seen as unresolved within traditional instrumental approaches, a reflexive turn emerges or in Beck's words reflexive modernisation (or 'world risk society').

The second element is a holistic approach in terms of scales, sectors and actors. As problem handling requires the possibility to transgress existing discursive and

[7] Metaphorically, single-loop learning is sometimes compared with a thermostat, which continuously makes adjustments based on incoming information (it is too hot), whereas double-loop learning is involved in searching for why a problem appears in the first place (why is it too hot). The second type of learning involves asking questions about the less apparent assumptions, values or governing factors underlying the system.

institutional boundaries (as emphasised in the first element above), it is essential that a variety of actors take part in the debate and search for solutions. The theory on reflexive governance, like EAM, acknowledges interdependencies across scales, sectors and arenas of governance, which in turn necessitates cooperation and dialogue among a variety of actors. The key question to address is whether there are arenas, forums and networks that allow for multi-actor, multi-scale as well as intersectoral collaboration and reflexivity. On a transnational regional level (e.g. the Baltic Sea region), this question and challenge is accentuated even more because institutional structures are traditionally developed within the system of sovereign nation states.

The theory of reflexive governance moreover has clear similarities with EAM and with scholarly writing about adaptive (co-)management in terms of the endorsement of a learning and dynamic view on governance (e.g. Hahn et al. 2006; Olsson et al. 2004), as well as regarding the important role of multiple actors in goal formulation, knowledge production and decision-making. While adaptive co-management primarily focuses on ecosystems, particularly at the local level, reflexive governance highlights the role played by social critiques at various levels. Social critique (or social reflexivity; cf. Beck 1992, 2006) plays a potentially constructive role for social and institutional learning. For individuals, organisations and institutions to be able to scrutinise themselves, they need to be open to the public confronting them about how their practices contribute to the production and reproduction of problematic circumstances. The reflexive governance perspective accordingly pays key attention to the importance of public debate and the monitoring role of civil society organisations and other actors.

The third element is recognition of path dependency and incremental policy-making. Acknowledgement of path dependency warrants, in turn, a consideration of incremental policy-making and step-by-step transformation. Radical and immediate shifts towards EAM in the complete sense are infeasible. In any governance arrangement, there are a number of exogenous and endogenous factors that serve to reproduce existing institutional structures (e.g. economic incentives, existing institutional boundaries, discourses, vested interests, habits, cognitive limitations and bounded rationality): "the existing structure fight back" (Grin 2006: 74). Thus, social-ecological transformation is typically path dependent, and existing physical and institutional structures significantly shape ambitions to work towards sustainable development. However, forward-looking and careful efforts to avoid and escape lock-in effects are possible even in an incremental approach to policy reform, Kemp and Loorbach (2006) argue. Moreover, path dependency also means that continuing along existing paths sometimes is preferable to keeping too many doors open as prior investments, learning and experiences then can be made use of, making capacity growth cumulative.

7.5 Analysis of Reflexivity in Baltic Sea Environmental Governance

7.5.1 Acknowledgement of Uncertainty and Ambivalence

Acknowledgement of uncertainty and ambiguity includes a willingness to actively explore uncertainty and ambiguity and develop preparedness for unintended consequences. Unintended consequences are impossible to eliminate completely and therefore have to be prepared for. This preparedness has to be both discursive and institutional. EAM and reflexive governance acknowledge an open-ended, experimental and learning-oriented approach, 'in the course of modulating ongoing developments, rather than towards complete knowledge and maximisation of control' (Voss and Kemp 2006: 7). Acknowledging uncertainty is thus something more than just a matter of trying to close information or knowledge gaps (Lidskog et al. 2009: 131–2). Reflexive governance entails an awareness that current modes of risk governance potentially could produce new side effects, new risks. Institutional and technological structures need to be forward-looking and adaptive, allow for trial-and-error learning and experiment with new innovations (Grin 2006; Kemp and Loorbach 2006).

While interviewees in the five case studies in general acknowledged the complexities and uncertainties surrounding environmental risks (regarding causes and effects), they seldom used tools to systematically take into account uncertainties and ambiguity; particularly those that stem from combined risks (e.g. cocktail effects regarding chemical risks). Our analysis showed that environmental risk governance of the Baltic Sea, in general, does not yet acknowledge these combined risks adequately. Risk assessment tends to go on as usual in the various sectors (Hassler et al. 2011, 2013). In none of the cases did we find any specific mechanisms, strategies or guidelines for systematically dealing with uncertainty and ambivalence. While the precautionary principle is increasingly invoked as a means to cope with uncertainty, actors rarely reflect upon how this principle should be implemented in practice. Having said that, what the case studies also show is that there is at least a growing recognition among scientists and other stakeholders that science alone cannot provide answers to all questions (see also Linke et al. 2016).

7.5.2 Holistic Approach in Terms of Scales, Sectors and Actors

A holistic approach entails preparedness for (1) multi-scale, (2) multi-sector and (3) multi-actor approaches to problem handling.

7.5.2.1 Multi-scale

First, multi-scale aspects include the possible need to shift scales – rescale – and deal with problems that go beyond existing geographical boundaries. There is no universal model here. Functional spaces may differ between risk areas, which can lead to considerable problems for the governance system. 'In short, the interaction space needs to be congruent with the problem space' (Voss et al. 2006a: 427). It is more the rule than the exception that environmental risks are transnational in character, implying that the robust handling of such risks necessitates the parallel development of new kinds of political and regulatory arenas, frameworks, actors and interactions that better correspond with the problem structures (cf. Lidskog et al. 2009). The cases of overfishing and eutrophication are most clearly regional problems and *collective bads*, which require regional collaboration. Yet, existing regulatory failures in these areas (see relevant chapters in the book) are not only due to the fact that the Baltic Sea countries are different. They are also due to the fact that these countries are affected differently, which poses more challenges to joint collaboration at the regional level. Depending on the size of costs and benefits (increased/improved ecosystem services), governments tend to adopt somewhat different strategies and have different ambitions when it comes to implementation, monitoring and enforcement.

Key EU directives and strategies may help to establish a regional approach to problem handling. We see in the cases of agriculture (Karlsson et al. 2016) and fishing (Sellke et al. 2016) a comparatively integrated approach at the European level to problem handling. A platform exists for the search for improved environmental risk governance. The EU can thus potentially help to move beyond a merely national view on what are regional problems. The fact that Baltic Sea countries have joined the EU in different enlargement phases and thus have had unequal amounts of time to adapt to EU regulations and practices, and also have very different political and economic backgrounds, makes it complicated to assess the full impact of the EU. One stated problem is that the relatively centralised EU regulatory framework has prevented sustainable regional solutions in these two sectors, although there are signs of regionalisation in the case of overfishing.

The cases of invasive alien species, oil spills and chemical pollution (see relevant Smolarz et al. 2016; Karlsson et al. 2016; Hassler 2016) exemplify regional risks relating to global flows of raw material, products and people. The management of these risks is more than in the other cases dependent on linking regional governance to global frameworks. For example, there are some elements in the European chemical regulation that serve to combine regional governance with a global kind of horizontal governance (along global product chains). Importers (public and private) need to be better aware of the chemical contents in products, which in turn means that they have to facilitate steering mechanisms, dialogue and information sharing with suppliers in other parts of the world (Boström et al. 2012; Boström and Karlsson 2013). However, this kind of combination of regional/global as well as vertical/horizontal aspects of governance is only in an embryonic stage. There is still a strong unmet demand for institutional/international cooperation.

The International Maritime Organization (IMO) plays a crucial role in the case of oil spills and appears to be a regulatory body that can relatively well foster and institutionalise such collaboration (Knudsen and Hassler 2011; see also Hassler 2016). The case study on IAS, in contrast, illustrated the existence of an embryonic global regulatory framework, the International Convention for the Control and Management of Ships' Ballast Water and Sediments (BWMC). This convention, however, has not yet (August 2015) entered into force.[8]

HELCOM has played an important role with regard to all these issues by establishing a political and regulatory space at the regional level and more recently through harmonising its regulations with EU regulations (Kern and Löffelsend 2004: 124). Although HELCOM lacks strong enforcement mechanisms, several interviewees we spoke to emphasised the many important roles that HELCOM plays. They said that HELCOM embodies an important part of the governance structure, as it is the only regional forum for negotiations where the entire BSR (Baltic Sea region) is represented, including Russia. It also plays an important role in linking Russia with the EU in debates and policy-making. HELCOM furthermore increasingly serves as an important arena for information sharing, negotiations and policy-making (e.g. Boström et al. 2015; VanDeever 2011), despite the parallel strengthening of EU competence in the Baltic Sea region.

HELCOM helps to coordinate countries in areas where there are many, partly overlapping and conflicting regulations (eutrophication, chemical case). It provides data and action plans (BSAP) and identifies hotspots. Based on such information, national NGOs as well as public authorities can then promote the implementation of programmes, projects and decisions made by HELCOM. HELCOM can thus function as an important reference point – a cognitive authority rather than rule-setting authority – for nationally based campaign organisations. HELCOM can also take a proactive approach in issue areas that are poorly developed.

7.5.2.2 Multi-sector

Second, the multi-sector notion acknowledges interdependencies across various sectors and arenas of governance, a key concern according to the EAM perspective. While we see clear paths towards institutionalisation of regional organisations and collaboration within sectors, there is less evidence of institutionalised collaboration *across* sectors. Governance so far has coped with this complexity primarily on the discursive level, i.e. by many authorities endorsing EAM, but not translating it into reformed practices. All case studies reveal substantial difficulties to translate this principle into practice. Each issue area has been traditionally handled separately by regulations and institutions, both nationally and within the EU. Governance relies on various assessment and management techniques (such as the total allowable catch system in the overfishing case) that are difficult to reconcile with EAM.

[8]The status of IMO Conventions can be checked at http://www.imo.org/en/About/Conventions/StatusOfConventions/Pages/Default.aspx

However, the potential role of HELCOM could be highlighted. It is not only a body that contributes to the carving out of a regional regulatory and political space. It is also a multi-issue organisation. According to our informants in HELCOM, different sections and working groups usually work separately. Yet, it is not hard to imagine that the increasing focus on EAM could pave the way for more systematic cross-sector collaboration (e.g. committees, working groups) within the organisation. A limit, on the other hand, is that while HELCOM may be able to achieve integration among environmental issues, it may face difficulties to reach out to other sectoral interests (agriculture, fishery).

It is likely that day-to-day concrete management of environmental risks in the Baltic Sea, as in other regions, will continue to be undertaken within sectoral administrations. The need for expertise, experience, hardware and specialised institutional structures makes this almost unavoidable. However, the initiatives with HELCOM forums – HELCOM AGRI/ENV Forum and FISH/ENV Forum – where public administration representatives of environmental and natural resource use sectors regularly meet to discuss the implementation of BSAP are interesting possibilities and could prove to be valuable in bridging sectors.

7.5.2.3 Multi-actor

Third, the multi-actor notion stresses the need to integrate the plurality of values and interests by using democratic procedures. Moreover, the need for transdisciplinary knowledge production, as well as co-production of knowledge among various types of knowledge holders needs to be acknowledged (Kemp and Loorbach 2006). Stakeholder inclusion is generally seen as a necessary element in reflexive governance (see also Jönsson et al. 2016). Different actors can shed additional light on common problems. Various sector organisations, NGOs and citizens can engage in social critique and environmental monitoring (Mol 2008), which potentially provide essential input to reflexive governance. These actors make observations, undertake collections and compile and interpret data – in parallel to governmental bodies and scientists doing the same – and then engage in framing and politicising of issues. It is important that monitoring is diffused among a plurality of actors, since this widens the way environmental risks are perceived, measured or estimated, which in turn can increase likelihoods that less appropriate methods and practices are weeded out. The simple fact that similar risks are approached from somewhat different angles can stimulate discussions on methods and methodology and contribute to refinement of environmental risk monitoring.

The case studies definitely show there are lots of stakeholders involved in governance and environmental monitoring. The Baltic Sea region scores relatively well as a marine region in terms of development of transnational networks (Joas et al. 2008; Kern and Löffelsend 2004). The interplay between HELCOM and non-state actors has been expanded (Van Deever 2011). HELCOM has undergone an attitude change and nowadays allow for non-state actors to take part as observers and even organises stakeholder conferences (Dreyer et al. 2011; Linke et al. 2016). Recent innovative forums include the Regional Advisory Councils in the fishing sector (see also Dreyer et al. 2014).

Nonetheless, there is a striking lack of *institutionalised structures* for stakeholder participation and communication, particularly at the *regional level*. And in general, there is a fairly big gap between ideal scenarios and practice when it comes to broad stakeholder participation and communication of environmental risk governance of the Baltic Sea (including the EU level) (Dreyer et al. 2011, 2014; Jönsson et al. 2016; Linke et al. 2016). From the case studies, we found scant evidence of real reflection on and practice of participatory and communication activities. Forms for horizontal interaction among groups are generally lacking (exceptions are umbrella NGOs such as Coalition Clean Baltic). HELCOM and other governance arrangements are only beginning to foster such horizontal interaction. Available forms of stakeholder interaction are fragmented and only allow for some restricted inclusion of various stakeholders, as within the EU WFD. The institutional design of the RACs presents one of the most innovative and interesting models. Yet, the RACs only apply to the risk management phase, not earlier phases during, for example, risk assessment and risk appraisal when the problem framing mainly occurs (see Dreyer et al. 2014; Sellke et al. 2010). Furthermore, there is a dominant functionalist view of participation in which stakeholders are at the service of policy. By interpreting the data from the cases studies, it is clear that stakeholders are supposed to improve implementation and provide useful knowledge to policy, but not expected to express their values or provide criticism based on reflexive thinking (social reflexivity in Beck's understanding). If the risk assessment arrangement relies on a narrow scientific-technical framing, which was particularly apparent in the eutrophication case, the governance framework in effect largely excludes stakeholders' knowledge and experiences and undervalues their potential input.

However, the critical issue is not merely that of increasing inclusion. While addressing tensions between inclusion of stakeholders and decision efficiency, Voss et al. (2006b) make an important observation by referring to an 'efficacy paradox', between 'opening up' for the inclusion of more actors and 'closing down' for decision-making. The more actors that become involved, the trickier the decision-making process is likely to be. This paradox, or tension, cannot be eliminated but needs to be recognised and somehow organisationally balanced. Although vast networks of international substate, city networks or lobby organisations exist within the BSR (see Fig. 3.1. in Haahti et al. 2010: 23), their participation is to a large extent predetermined by the character of the regulatory framework, making their role more indirect, i.e. attending meetings and participating within, for example, HELCOM according to their 'observer' status. But, due to the vastness of the NGO networks, competition among networks, for example, for funding and hence domination, this has in some cases resulted in overlapping and conflicting agendas. A perception that there are too many actors engaged is likely to emerge if there are a lack of clear forms and structures for stakeholder input as well as a lack of rules and guidelines that define the roles and relationships among stakeholders. Absence of structures for stakeholder participation and communication corresponds with a lack of coordination among actors. Furthermore, it should be carefully noted that reflexivity within the stakeholder participation process will suffer if the dominant participants guard too close their specific interests. For example, it was illustrated in the case studies on eutrophication and overfishing that sector interests (agriculture and fisheries, respectively) made reflexivity, open-ended discussions and long-term strategic thinking problematic. In other areas, where stakeholders do not have such strong sectoral interests to defend, progress is more likely.

Finally, it is important to acknowledge that a plurality of actors each can take their own initiatives and a central place in the development of rules, policies and guidelines. Examples of such initiatives include voluntary agreements between public and private actors, codes of conduct that companies use in interaction with suppliers, eco-certification and labelling and guidelines for sustainable procurement and other practices. The perspective of reflexive governance stresses that this role is not limited to governmental actors (on either the national or international level). Often there is a need for voluntary approaches initiated by non-governmental actors (either for profit or not for profit) when the regulatory space is diffuse and relatively unregulated. Some actors take initiatives and suggest templates (which the inter-governmental actor HELCOM also does). It is interesting to note that businesses take their own initiatives and engage in risk assessments and research in order to develop technical solutions. This cannot just be dismissed as greenwashing but could indicate also instances of reflexivity, as it occurs as a response to the slow implementation of existing regulations. In the case study of IAS, business actors took own rule-making initiatives, and it was reported that companies encouraged states to ratify the convention. As long as shipping companies can continue their business in a technically and economically feasible way, there is no self-evident reason for them to counteract environmentally sound regulation. As Young observes (2009: 25), 'corporate actors are frequently more concerned with the development of stable rules and a uniform and predictable regulatory environment than with the exact content of the resultant governance systems'. Voluntary rule-making can sometimes, in effect, appear as strict and compelling as mandatory regulation (or even more strict, because there can be considerable interpretative flexibility and room for manoeuvre within measures that are defined as mandatory). Delmas (2009) explains this observation by referring to March and Olsen's well-known distinction between 'the logic of consequences' and 'the logic of appropriateness'. The first one relates to calculation of positive economic or noneconomic benefits for individual and collective actors, whereas the latter refers to when actors respond to cognitive or normative pressures in order to establish legitimacy or avoid their reputations from being damaged. Soft approaches can very well relate to both these types of logics. A combination of voluntary rule-making and mandatory regulation is also necessary in the cases when global flows of raw material and products relate to the regional risks and where effective global conventions are absent (Boström and Karlsson 2013). The mix of hard and soft approaches should accordingly be seen as context dependent and could thus be adjusted depending on the scale and scope of the problem structure.

7.5.3 *Path Dependency and Incremental Policy-Making*

The reflexive governance perspective underscores the importance of taking path dependency seriously. Arguably, considerable reflection is required to determine when incremental, step-by-step transformation is appropriate rather than searching for utopian policies, or policies that might seem optimal but unfeasible for various reasons (Grin 2006; Kemp and Loorbach 2006), and when it is wiser to continue with prior investments and already selected paths. While path dependency metaphorically denotes a barrier to change, one could also see it in more positive light, namely, as 'pathways'. Are there any windows of opportunity to overcome lock-in effects and induce new pathways, which on the one hand will create a level of path dependency in the future, but on the other could be seen as a more favourable path? The search for such pathways requires reflexivity.

Some of the dominant intergovernmental organisations in relation to environmental risks in the Baltic Sea have comparably long histories. HELCOM is experiencing its 40-year anniversary, IMO has already turned 60, and the International Council for the Exploration of the Sea (ICES) has a history that goes back all the way to 1902. Although they all have evolved over time and affected by external and internal change pressures, large organisations tend to change only slowly. This is natural as procedures, practices and expectations become increasingly established and part of the organisation's identity. Moreover, as long as most of the staff remains part of the organisation for a long time, change is often circumscribed because of existing expertise and the perspectives adopted by leading experts and administrators within the organisation. This may sometimes create organisational inertia and a limited ability to adapt to changing societal expectations (Ahrne and Papakostas 2002). However, the experiences, know-how and expertise that have been built up in the organisations over time can also be important assets. The authority and influence exerted by these organisations are often the result of well-established expertise (Barnett and Finnemore 2004).

It is important to note that both aspects – potential organisational inertia and established expertise – contribute to path dependency, namely, difficulties in changing perspectives such as moving from a traditional reductionist and sector scientific perspective to the ecosystem approach to management. From a policy perspective, this needs to be carefully assessed in relation to the creation of space for reflexivity. Looking, for example, at the emerging attention being given to fishery and eutrophication issues by HELCOM, it is possible that a strengthened role for HELCOM in these areas will increase the weight given to traditional scientific practices (single-loop learning), as it is with these practices that the organisation has its strongest merits. On the other hand, it is also possible that HELCOM entering into less chartered areas, creating new committees and working groups end engaging new types of experts will create opportunities for the adoption of new perspectives, perspectives based on contemporary thinking on sustainability and holistic ecosystem approaches (double-loop learning). Compare this with the contemporary trend towards EU regionalisation in fisheries where new bodies (RACs) have been set up according to the perceived importance of stakeholder participation, utilisation of different forms of knowledge and decentralisation. Here, the creation of new organisational bodies creates space for new thinking and breaking up of organisational inertia, while at the same time previously built up experience and knowledge to some extent may be lost (see also Hassler et al. 2013 about recent and current developments in ICES, in which organisational change opened up possibilities for increasing reflexivity).

It should be noted that not only organisations can be trapped in various forms of path dependency, but individual sectors as well, not least so in ways of thinking and perspectives adopted. It is possible that the fishery sector in the EU to a considerable extent suffers from what may be called TAC path dependency (Hassler et al. 2013; Sellke et al. 2010, 2016). Reflection on how to better manage Baltic Sea fish stocks is still hampered by traditional thinking on maximum sustainable yield (MSY) and TACs that makes elaboration of concrete management plans based on EAM slow to emerge. Somewhat similarly, eutrophication caused by agriculture seems to be trapped in a reductionist thinking regarding natural resource management, where the EU/CAP system often sets the agenda for discussions and circumscribes what is brought to the table (Karlsson et al. 2016; Haahti et al. 2010).

The breaking loose from negative aspects of path dependency is difficult. Established organisations are often inert and slow to change, historic investments create vested interests, and expert and public deliberation is often dominated by prior discourses. However, one way to move forward can be to establish new institutions for discussion and reflection and stimulate new projects or programmes in established organisations by providing ear-marked funding and support to organisations entering new areas. One example are HELCOM's forums for cross-sector discussions between on the one hand the agriculture and environment sectors and on the other the fisheries and environment sectors. An argument could be made that this 'institutional redundancy' where several organisations deal with the same risk sector – but based on different perspectives and approaches – could, in fact, in the long run be better than too much streamlining and specialisation. Considering the tremendous difficulties in reforming CAP and CFP to fall more in line with sustainability ambitions, different ways to find new perspectives and approaches are without question badly needed.

7.6 Towards EAM and Reflexive Governance?

The contemporary system of environmental governance in the Baltic Sea predominately emphasises single species and tightly circumscribed risk assessment procedures that tend to exclude all other expertise and even formal experts from decision-making and management. The rationale for new forms of governance is that the traditional governance mode is ill-equipped to deal with the complexity of environmental governance in a holistic manner, namely, giving systematic attention to uncertainty, plurality of values, ambiguity and limited knowledge.

Substantial amounts of uncertainty necessitate risk assessments drastically different than those done in the traditional way. Governance modes need to be developed that focus more on how to handle uncertainties and induce reflexivity on existing paths and possible futures, rather than on finding exact predictions on hazardous effects and outcomes. Moreover, all five risk areas are characterised by significant amounts of ecological and social complexity. These complexities make governance challenging. Issues such as regulatory overlap and overload, antagonistic institutional interactions, lagging enforceability, unbalanced inclusion of stakeholders, slow regionalisation, poor risk communication and several others are urgently needed to be addressed in all five areas addressed in this book. A particularly crucial issue discussed in this chapter is the, oftentimes cemented, sectorisation that prevents interaction and learning across sectors. Despite the variation in solutions that most likely is needed, learning across risk areas can potentially improve over-all environmental safety levels and benefit long-term sustainability. While we found very little of such cross-sector interaction, early attempts in organisations such as HELCOM and ICES do indicate fruitful pathways and signs of emerging reflexivity (Hassler et al. 2013).

Although comprehensive regulatory frameworks for managing the Baltic Sea are mostly in place, existing institutional structures accordingly lack a holistic approach to management, as the structures are often sector driven and based on sectoral governance that often excludes cross-sector cooperation and reflection. In addition, different regulatory frameworks contradict each other and constitute a confusing totality. The results from the case studies show that the traditional emphasis on single species and tightly circumscribed risk assessments that rely on narrow scientific-technical framings still dominate, despite a broader and increasing recognition of the problems with these narrow views. There are many stakeholders around demanding attention. However, the structures that would facilitate fruitful coordination and deliberations among them are few, limiting and vague (see also Jönsson et al. 2016). Existing institutions are thus inclined to exclude co-management while mechanisms of path dependency are likely to force actors to stick with 'business as usual'. Managing the Baltic Sea entails substantial amounts of uncertainty due to ecological and social complexity, but there is an apparent lack of scientific interdisciplinary integration, linking natural and socio-economic science in assessing the complexity of risk sources, as existing assessment is primarily based on natural science. The institutional structures also tend to overlook how to deal with scientific uncertainty, although there appears to be growing awareness among scientists and other stakeholders that science alone cannot provide all the answers.

The ability to address EAM is a matter of dealing with inert organisations, path dependency and decision-making systems based on a short-term focus. As we have stressed, the concept of reflexive governance points to the need for strategic thinking on how to build institutions and governance modes that are both forward- and backward-looking, which entails reflection – double-loop learning – on the very governance process itself including its conditions. Such reflexivity cannot be done and induced from one vantage point, but rather from a plurality of them. Problem handling requires the possibility to transgress existing discursive and institutional boundaries. A precondition for such transgression is that a variety of actors can take part in the debate and search for solutions.

The EAM concept has potential to facilitate such type of reflection, but only if social sustainability issues such as participation, justice, quality of life and social coherence are more pronounced (see Boström 2012 for a discussion regarding social sustainability). Innovative concepts such as EAM may not induce quick changes but could still have an important role to play in the long run. New concepts bring new ways to frame problems and solutions and can open up new spaces for interaction and dialogue among actors and consequently for mutual learning among them. Thus, we should not be too hasty in assessing or neglecting the potential of EAM to bring about change. EAM does, indeed, bring reflexive potential to criticise traditional risk assessment and management (such as the TAC machine) and can open up doors at the cognitive level. EAM helps actors to look for new 'pathways'.

However, concepts such as EAM are not sufficient to accomplish such tasks on their own. Institutional and organisational structures that provide interactive space and remove obstacles for such reflexivity also need to be in place. This is only beginning to happen in regional governance of the Baltic Sea. Institutional structures appear even more inert than cognitive structures. While many interviewees used the rhetoric of sustainability, ecosystems, uncertainty, precaution, holism, cocktail effects and so on, they were unable to translate new ideas into feasible operations. On the institutional level, a welcome development is, however, the increasing cross-border interaction and capacity building that have emerged from both above and below. Tighter interaction and capacity building among non-state actors, for instance, provide better conditions for constructive input to policy at this cross-border regional level. It should be remembered that input from non-state actors is crucial not only in the more instrumentalist/functionalist way (for the service of current governance) but also in the more reflexive/critical way (social reflexivity). The latter is so because the incessant push for change is arguably needed to prevent the current governance system from relapsing into business as usual while only rhetorically committing to EAM.

Acknowledgements This chapter draws on results from the research programme 'Environmental Risk Governance of the Baltic Sea' (2009–2015; Michael Gilek, programme coordinator) and involved research teams from Södertörn University in Sweden, Åbo Akademi University in Finland, Dialogik/Stuttgart University in Germany and Gdansk University in Poland. The funding came from the Foundation for Baltic and East European Studies and the European Community's Seventh Framework Programme (2007–2013) under grant agreement no. 217246 made with the joint Baltic Sea research and development programme BONUS, as well as from the German Federal Ministry of Education and Research (BMBF), the Swedish Environmental Protection Agency, the Swedish Research Council FORMAS, the Polish Ministry of Science and Higher Education and the Academy of Finland. We wish to express our warmest thanks to these institutions for enabling us to conduct this research, to all participants in the research programme that directly or indirectly provided useful input, to all informants who shared their experiences and to the two reviewers for their constructive comments on an earlier version of the chapter.

References

Ahrne G, Papakostas A (2002) Organisationer, samhälle och globalisering: tröghetens mekanismer och förnyelsens förutsättningar (in Swedish). Studentlitteratur AB, Lund

Arkema KK, Abramson SC, Dewsbury BM (2006) Marine ecosystem-based management: from characterization to implementation. Front Ecol Environ 4(10):525–532

Backer H, Leppänen JM, Brusendorff AC, Forsius K, Stankiewicz M, Mehtonen J, Pyhälä M, Laamanen M, Paulomäki H, Vlasov N, Haaranen T (2010) HELCOM Baltic Sea Action Plan – a regional programme of measures for the marine environment based on the ecosystem approach. Mar Pollut Bull 60:642–649

EU Balticnet (2012) Fact sheet, May 2012

Barnes C, McFadden KW (2007) Marine ecosystem approaches to management: challenges and lessons in the United States. Mar Policy 32:387–392

Barnett M, Finnemore MA (2004) Rules for the world. International organisations in global politics. Cornell University Press, London

Beck U (1992) Risk society. Towards a new modernity. Sage Publication, London

Beck U (1994) The reinvention of politics. Towards a theory of reflexive modernization. In: Beck U, Giddens A, Lash S (eds) Reflexive modernization. Politics, tradition and aesthetics in the modern social order. Polity Press, Cambridge

Beck U (2006) Reflexive governance: politics in the global risk society. In: Voss JP, Bauknecht D, Kemp R (eds) Reflexive governance for sustainable development. Edward Elgar, Cheltenham, pp 31–56

Berghöfer A, Wittmer H, Rauschmayer F (2008) Stakeholder participation in ecosystem-based approaches to fisheries management: a synthesis from European research projects. Mar Policy 32:243–253

Boström M (2012) A missing pillar? Challenges in theorizing and practicing social sustainability. Sustain: Sci Pract Policy 8(1):3–14

Boström M, Karlsson M (2013) Responsible procurement, complex product chains and the integration of vertical and horizontal governance. Environ Policy Gov 23(6):381–394

Boström M, Börjeson N, Gilek M, Jönsson AM, Karlsson M (2012) Responsible procurement and complex product chains: the case of chemical risks in textiles. J Environ Plan Man 55(1):95–111

Boström M, Rabe L, Rodela R (2015) Environmental non-governmental organizations and transnational collaboration in two macro-regional contexts: the Baltic Sea and Adriatic-Ionian Sea regions. Environ Polit 24(5):762–787.

BSR programme (2012) Executive summary. Available from: http://eu.baltic.net. Accessed 23 May 2012

Curtin R, Prellezo R (2010) Understanding marine ecosystem based management: a literature review. Mar Policy 34:821–830

Delmas MA (2009) Research opportunities in the area of governance for sustainable development. In: Delmas MA, Young O (eds) Governance for the environment. New perspectives. Cambridge University Press, Cambridge, pp 221–238

Dreyer M, Selke P, Jönsson AM, Boström M (2011) Structures and processes of stakeholder and public communication on Baltic Sea environmental risks. RISKGOV report, delivery number 10. Södertörn University, Huddinge, Available from: http://www.sh.se/riskgov

Dreyer M, Boström M, Jönsson AM (2014) Participatory deliberation, risk governance and management of the marine region in the EU. J Environ Policy Plan. doi:10.1080/1523908X.2013.866891

Galaz V, Olsson P, Hahn T, Folke C, Svedin U (2008) The problem of fit among biophysical systems, environmental and resource regimes, and broader governance systems: insights and emerging challenges. In: Young OR, King LA, Schröder H (eds) Institutions and environmental change – principal findings, applications, and research frontiers. MIT Press, Cambridge, pp 147–182

Gilek M, Linke S, Udovyk O, Karlsson M, Lundberg C, Smolarz K, Lemke P (2011) Interactions between risk assessment and risk management for environmental risks in the Baltic Sea. Deliverable 9 within the RISKGOV project. Södertörn University, Huddinge, Available from: http://www.sh.se/riskgov

Gilek M, Karlsson M, Linke S, Smolarz K (2016) Environmental governance of the Baltic Sea: identifying key challenges, research topics and analytical approaches. In: Gilek M et al (eds) Environmental governance of the Baltic Sea. Springer, Dordrecht

Grin J (2006) Reflexive modernisation as a governance issue, or: designing and shaping restructuration. In: Voss JP, Bauknecht D, Kemp R (eds) Reflexive governance for sustainable development. Edward Elgar, Cheltenham, pp 57–81

Grumbine RE (1994) What is ecosystem management? Conserv Biol 8(1):27–38

Grumbine RE (1997) Reflections on 'what is ecosystem management?'. Conserv Biol 11(1):41–47

Haathi BM, Hedenström E, Linke S, Lundberg C, Reisner G, Wanamo M (2010) Case study report: eutrophication. RISKGOV report. Södertörn University, Huddinge, Available from: http://www.sh.se/riskgov

Hahn T, Olsson P, Folke C, Johansson K (2006) Trust-building, knowledge generation and organizational innovations: the role of a bridging organization for adaptive comanagement of a wetland landscape around Kristianstad, Sweden. Hum Ecol 34(4):573–592

Hassler B (2003a) Protecting the Baltic Sea – the Helsinki convention and national interests. In: Schram O, Thommessen OB (eds) Yearbook of international co-operation on environment and development 2003–2004. Earthscan, London

Hassler B (2003b) Science and politics of foreign aid – Swedish environmental support to the Baltic states. Kluwer Academic Publishers, Dordrecht

Hassler B (2016) Oil spills from shipping: a case study of the governance of accidental hazards and intentional pollution in the Baltic Sea. In: Gilek M et al (eds) Environmental governance of the Baltic Sea. Springer, Dordrecht

Hassler B, Söderström S, Lepoša N (2010) Marine transportations in the Baltic Sea area. RISKGOV report, delivery number 6. Södertörn University, Huddinge, Available from: http://www.sh.se/riskgov

Hassler B, Boström M, Grönholm S, Kern K (2011) Environmental risk governance in the Baltic Sea – a comparison among five key areas. RISKGOV report, delivery number 8. Södertörn University, Huddinge, Available from: http://www.sh.se/riskgov

Hassler B, Boström M, Grönholm S (2013) Towards an ecosystem approach to management in regional marine governance? The Baltic Sea context. J Environ Policy Plan 15(2):225–245

Hjorth R (1992) Building international institutions for environmental protection: the case of Baltic Sea environmental cooperation. Doctoral thesis, Linköping Studies in Arts and Science, Linköping University

Joas M, Jahn D, Kern K (2008) Governance in the Baltic Sea region: balancing between states, cities and peoples. In: Joas M, Jahn D, Kern K (eds) Governing a common sea: environmental policies in the Baltic Sea region. Earthscan, London, pp 3–18

Jönsson AM, Boström M, Dreyer M, Söderström S (2016) Risk communication and the role of the public: towards inclusive environmental governance of the Baltic Sea? In: Gilek M et al (eds) Environmental governance of the Baltic Sea. Springer, Dordrecht
Karlsson M, Gilek M (2016) Governance of chemicals in the Baltic Sea region: a study of three generations of hazardous substances. In: Gilek M et al (eds) Environmental governance of the Baltic Sea. Springer, Dordrecht
Karlsson M, Gilek M, Lundberg C (2016) Eutrophication and the ecosystem approach to management: a case study of Baltic Sea environmental governance. In: Gilek M et al (eds) Environmental governance of the Baltic Sea. Springer, Dordrecht
Kemp R, Loorbach D (2006) Transition management: a reflexive governance approach. In: Voss JP, Bauknecht D, Kemp R (eds) Reflexive governance for sustainable development. Edward Elgar, Cheltenham, pp 103–130
Kern K (2011) Governance for sustainable development in the Baltic Sea region. J Baltic Stud 42(1):67–81
Kern K, Löffelsend T (2004) Sustainable development in the Baltic Sea region. Governance beyond the nation state. Local Environ 9(5):451–467
Knudsen OF, Hassler B (2011) IMO legislation and its implementation: accident risk, vessel deficiencies and national administrative practices. Mar Policy 35(2):201–207
Lemke P, Smolarz K, Zgrundo A, Wolowicz M (2010) Biodiversity with regard to alien species in the Baltic Sea region. RISKGOV report to BONUS EEIG Programme. University of Gdansk, Institute of Oceanography, Gdynia, Available from: http://www.sh.se/riskgov
Lidskog R, Soneryd L, Uggla Y (2009) Transboundary risk governance. Earthscan, London
Linke S, Gilek M, Karlsson M (2016) Science-policy interfaces in Baltic Sea environmental governance: towards regional cooperation and management of uncertainty? In: Gilek M et al (eds) Environmental governance of the Baltic Sea. Springer, Dordrecht
Mol A (2008) Environmental reform in the information age. The contours of informational governance. Cambridge University Press, New York
Murawski AM (2007) Ten myths concerning ecosystem approaches to marine resources management. Mar Policy 31:681–690
Oberthür S, Gehring T (eds) (2006) Institutional interaction in global environmental governance: synergy and conflict among international and EU policies. MIT Press, Cambridge, MA
Olsson P, Folke C, Berkes F (2004) Adaptive comanagement for building resilience in social–ecological systems. Environ Manag 34(1):75–90
Österblom H, Gårdmark A, Bergström L, Müller-Karulis B, Folke C, Lindegren M, Casini M, Olsson P, Diekmann R, Blenckner T, Humborg C, Möllmann C (2010) Making the ecosystem approach operational – can regime shifts in ecological and governance systems facilitate the transition? Mar Policy 34:1290–1299
Renn O (2008) Risk governance: coping with uncertainty in a complex world. Earthscan, London
Sellke P, Dreyer M, Renn O (2010) Fisheries: a case study of environmental risk governance in the Baltic Sea. RISKGOV report, delivery number 3. Södertörn University, Huddinge, Available from: http://www.sh.se/riskgov
Sellke P, Dreyer M, Linke S (2016) Fisheries: a case study of Baltic Sea environmental governance. In: Gilek M et al (eds) Environmental governance of the Baltic Sea. Springer, Dordrecht
Smolarz K, Biskup P, Zgrundo A (2016) Biological invasions: a case study of Baltic Sea environmental governance. In: Gilek M et al (eds) Environmental governance of the Baltic Sea. Springer, Dordrecht
Stirling A (2009) Participation, precaution and reflexive governance for sustainable development. In: Adger WN, Jordan A (eds) Governing sustainability. Cambridge University Press, New York
Stokke OS (2001) Conclusions. In: Stokke OS (ed) Governing high seas fisheries: the interplay of global and regional regimes. Oxford University Press, Oxford, pp 329–360
Tallis H, Levin PS, Ruckelshaus M, Lester SE, McLeod KL, Fluharty DL, Halpern BS (2010) The many faces of ecosystem-based management: making the process work in real places. Mar Policy 34:340–348

Udovyk O, Rabilloud L, Gilek M, Karlsson M (2010) Hazardous substances: a case study of environmental risk governance in the Baltic Sea region. RISKGOV report to BONUS EEIG programme. Södertörn University, Sweden

VanDeever SD (2011) Networked Baltic environmental cooperation. J Baltic Stud 42(1):37–55

Voss JP, Kemp R (2006) Sustainability and reflexive governance: introduction. In: Voss JP, Bauknecht D, Kemp R (eds) Reflexive governance for sustainable development. Edward Elgar, Cheltenham, pp 3–28

Voss JP, Bauknecht D, Kemp R (eds) (2006a) Reflexive governance for sustainable development. Edward Elgar, Cheltenham

Voss JP, Kemp R, Bauknecht D (2006b) Reflexive governance: a view on an emerging path. In: Voss JP, Bauknecht D, Kemp R (eds) Reflexive governance for sustainable development. Edward Elgar, Cheltenham, pp 419–437

Walker G, Shove E (2007) Ambivalence, sustainability, and the governance of socio-technical transitions. J Environ Policy Plan 9:213–225

Young O (2002) The institutional dimensions of environmental change: fit, interplay, and scale. MIT Press, Cambridge, MA

Young O (2009) Governance for sustainable development in a world of rising interdependencies. In: Delmas MA, Young O (eds) Governance for the environment. New perspectives. Cambridge University Press, Cambridge

Chapter 8
Science-Policy Interfaces in Baltic Sea Environmental Governance: Towards Regional Cooperation and Management of Uncertainty?

Sebastian Linke, Michael Gilek, and Mikael Karlsson

Abstract This chapter investigates and compares the interactions between science and policy (risk assessments and risk management) in five cases of environmental governance of the Baltic Sea: eutrophication, fisheries, invasive alien species, chemical pollution and oil discharges. An efficient interplay between science and policy is important for successful environmental governance, which applies particularly to the Baltic Sea where all five risks pose serious threats to environmental, social and economic aspects of sustainability. We use science-policy theory and an analytical framework based on a categorisation of relevant management responses linked to different states of incomplete knowledge (risk, uncertainty, ambiguity, ignorance) to investigate two main characteristics of science-policy interfaces: (1) organisational structures and (2) procedural aspects of managing scientific uncertainties and stakeholder disagreements. The analyses reveal differences and similarities in institutional and organisational designs of the respective assessment-management interactions, as well as in terms of how scientific uncertainties, stakeholder disagreements and sociopolitical ambiguities are addressed. All the five science-policy interfaces expose science-based management approaches that commonly are not able to cope sufficiently well with the complexities, uncertainties and ambiguities at hand. Based on our cross-case analyses, we conclude by recommending five key aspects that need

S. Linke (✉)
Department of Philosophy, Linguistics and Theory of Science, University of Gothenburg, Box 200, 405 30 Göteborg, Sweden
e-mail: sebastian.linke@gu.se

M. Gilek • M. Karlsson
School of Natural Sciences, Technology and Environmental Studies, Södertörn University, 14189 Huddinge, Sweden
e-mail: michael.gilek@sh.se; mikael.karlsson@2050.se

M. Gilek et al. (eds.), *Environmental Governance of the Baltic Sea*, MARE Publication Series 10, DOI 10.1007/978-3-319-27006-7_8

to be addressed to improve science-policy interactions in Baltic Sea environmental governance: (*1*) more adaptive organisational structures in terms of time, context and place dependency, (*2*) increased knowledge integrations, (*3*) a more careful consideration of stakeholder participation and deliberation, (*4*) better management of uncertainty and disagreements and (*5*) increased transparency and reflection in the communication of science-policy processes.

Keywords Science-policy interactions • Marine policy • Post-normal science • Uncertainty management • Stakeholder participation

8.1 Introduction

The five cases of environmental governance studied in this project have been identified as key large-scale environmental problems and risks in the Baltic Sea (see Gilek et al. 2016; HELCOM 2010): fisheries, eutrophication, invasive alien species, chemical pollution and oil discharges linked to marine transportation. However, as revealed throughout the previous chapters, they differ substantially in terms of the complexity of risk sources, the available knowledge and the uncertainties connected to assessing environmental effects for advising decision-making, as well as with respect to the degrees of ambiguity and sociopolitical controversy involved in policy and management (Gilek et al. 2011).

In this chapter, we investigate the interactions between risk assessment (science) and risk management (policy) in the five different cases. We analyse and compare these interactions using a theoretical framework on science-policy interfaces described in Sect. 8.2. Specifically, we study how organisational structures and processes of science-policy interactions adapt to key challenges of science and management in environmental governance by focusing on different forms of uncertainty, as well as on stakeholder conflicts and disagreements involved in science and/versus policy in the five cases. We also trace the respective management reactions to these challenges in each of the cases using a typology of different kinds of incomplete knowledge and their consequences for management responses as described below (Sect. 8.2.2). Through this comparative study of science-policy interfaces across the five cases, we point out institutional and procedural hindrances, challenges and prospects for improving science-policy interactions for a more effective and sustainable environmental governance of the Baltic Sea. Following the discussion on our theoretical framework, we present the results of our analysis of the five science-policy interfaces (Sect. 8.3). The two sections thereafter discuss the outcomes of the study (Sect. 8.4) and provide conclusions and recommendations (Sect. 8.5), respectively.

8.2 Theoretical Context

8.2.1 *Science-Policy Theory and the Ecosystem Approach to Management*

Science-based advice is universally regarded and used as a primary trustworthy basis for environmental management and decision-making. However, at the same time, in many areas, there is an ambiguity and increasing concern about the sole dependence on expertise from the (natural) sciences, which often acts to the detriment of sufficient consideration of other knowledge claims, stakeholder perspectives and values held by actors such as NGOs, citizens or business people. Bijker et al. (2009: 1) have called this phenomenon the paradox of scientific authority and asked the question of 'how can scientific advice be effective and influential in an age in which the status of science and/or scientists seems to be as low as it has ever been?' One reason for this paradox lies in the fact that science becomes politicised whenever it is called into a political context (Weingart 1999) and is hence subject to constraints, i.e. rules, norms and evaluation criteria, other than those set by the scientific community. A basic challenge is that the demarcation that exists between the spheres of science and politics falls apart in such contexts, which consequently leads to concerns and potential conflicts about the legitimate role of science and its relation to policy. Both the role of science-based advice and the political decisions based on it may therefore become contested with respect to credibility, legitimacy and accountability (cf. Cash et al. 2003).

In management practice, a distinction is often made between a science system (representing factual knowledge claims) and a social system (representing political, business and other concerns of public life) – a division that is inscribed in the institutional design of most policy systems in modern societies. Science is responsible for providing the best available knowledge in terms of a presumed value-free and objective input to political decision-making, which is accordingly seen as most rational and democratically legitimated (Funtowicz and Strand 2007; Wilson 2009). This idealised model of interaction between science and policy relies on what has been called the 'ideal causal chain' of implementing scientific knowledge in policy processes (Fig. 8.1; Gezelius 2008).

The ecosystem approach to management (EAM) is seen as a necessary and idealistic approach for managing marine resources and other environmental issues with regard to all five areas of environment and risks analysed here (Backer et al. 2010; Garcia et al. 2003; Hammer 2015). As a consequence, assessment and management

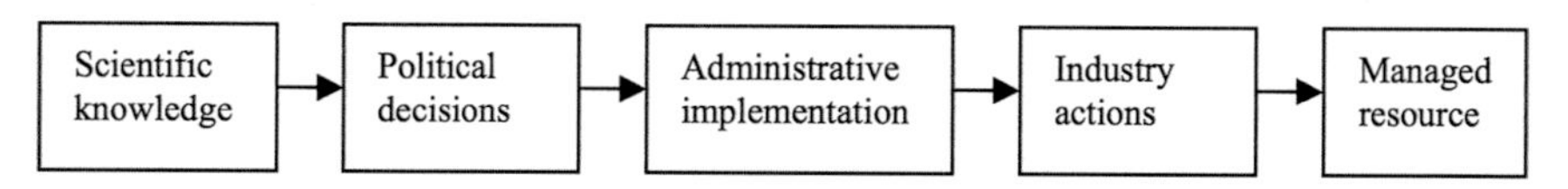

Fig. 8.1 The 'ideal causal chain' model of science input to management as, for example, described for EU fisheries by Gezelius (2008) (Reprinted with permission of Springer)

practices need to adapt to new ways of giving, using and implementing various sorts of advice. This shift towards EAM is furthermore in line with the EU's general principles of 'good governance' (COM 2001: 10) applied in various marine policies such as the Common Fisheries Policy (CFP), aiming for a 'broad involvement of stakeholders at all stages of the policy from conception to implementation' (EC 2002: 6).

Our study analyses the problems and challenges faced when a traditional conceptualisation of science-policy interfaces, such as the 'ideal causal chain' model, is applied to different cases of environmental governance in the Baltic Sea and suggests ways in which associated problems and challenges might be countered. Applying the EAM concept, with its aspirations to achieve sustainable use of ecosystems in line with place-based requirements and sensitivities of the socioecological system (cf. Boström et al. 2016), opens new opportunities for a more holistic approach to understand and design science-policy interfaces by taking both the natural system (represented via science) and the social system (sociopolitical aspects) into account (cf. Gilek et al. 2015). The widening perspective of EAM could hence contribute to avoiding some of the traditional pitfalls of narrowly assigned science-based management systems. One of these pitfalls is the strongly sector-based marine environmental governance of the Baltic Sea that is studied here.

Furthermore, all our case studies present strongly politicised domains of environmental governance (albeit with case-specific differences), where the boundaries between science and policy are continuously blurred and often debated. This implies that political and cultural values heavily influence scientific processes while science on the other hand strongly influences policy developments. This phenomenon is described as the 'co-production of science and policy' (Jasanoff 2004; Jasanoff and Wynne 1998) and highlights key questions about the roles and responsibilities of different actors such as scientists (and the science system) and other relevant stakeholders and policymakers in the interplay between assessment and management. In this 'co-produced' context, new and developed institutional structures and processes of interaction could act as 'boundary organisations' between science and policy that make environmental problems governable. As Lidskog (2014:3) states: 'By negotiating and renegotiating the boundaries between science and policy, environmental problems and their possible solutions are co-produced. Both science and policy are mobilised in order to solve a specific environmental problem'.

8.2.2 Analytical Framework and Methods

EAM emphasises the importance of two major aspects of environmental governance. It first highlights a *regional basis for management*. Second, it emphasises appropriate processes, methods and techniques for *dealing with uncertainties and disagreements* in the interaction between science and policymaking. Using an analytical framework focussing on science-policy interactions as outlined below, we investigate how these two aspects are dealt with in the five cases of environmental governance in the Baltic Sea.

Emphasising the importance of *organisational structures* and *procedural interactions* of the science-policy interface highlights if and how the challenges connected with assessment-management interactions differ between the five cases analysed in this study (as seen, e.g., when comparing fisheries and eutrophication – cf. Karlsson et al. 2016; Linke et al. 2014; Sellke et al. 2016). Stirling (2010: 1029) has argued with regard to the neglect of such relevant differences that an 'overly narrow focus on risk is an inadequate response to incomplete knowledge', because it makes the (necessarily simplified) science-based advice vulnerable to social interests, political manipulation and pressures from lobby groups. Stirling therefore suggests an 'opening up' of linear, scientific conceptions of the science-policy interface for more plural and situated understandings (Stirling 2008: 262). He also suggests it is necessary to take a more careful account of the nature of the knowledge at hand by saying that 'when the intrinsically plural, conditional nature of knowledge is recognised, I believe that science advice can become more rigorous, robust and democratically accountable' (Stirling 2010: 1029). In order to better adapt science-policy interactions to these insights, he has developed an 'uncertainty matrix' that differentiates between four different idealised states of incomplete knowledge (Fig. 8.2).

The formal state of *risk* (Fig. 8.2) is characterised by a comparatively high level of confidence in both the knowledge about possible outcomes as well as about their respective probabilities. It can thus be handled by traditional linear risk assessment-management procedures based on a straightforwardly applied scientific approach (as in Fig. 8.1). However, this is not the case in the three other cases of the matrix, namely, *uncertainty*, *ambiguity* and *ignorance*, which according to Stirling differ from the traditional risk categorisation.

Under the condition of scientific *uncertainty*, it is still feasible to characterise possible outcomes but the available information input (data) is too incomplete to assign specific probabilities (e.g. as often is argued for the enormous number of chemical pollutants in the environment). For such (uncertain) environmental issues, as Stirling (2007: 310) notes, 'the scientifically rigorous approach is therefore to acknowledge various possible interpretations'.

The condition of *ambiguity* is, on the other hand, not primarily characterised by problematic knowledge about probabilities (data input) but about the possible outcomes and contested interpretations and framings of the environmental issue. The management of various marine resources, such as commercial fish stocks, has been argued to belong to this type of an environmental issue (cf. Linke et al. 2014). For such cases, disagreements among disciplines and specialists may arise as a consequence of different integration of ecological, agronomic, safety or socio-economic criteria of harm. Therefore, the application of a traditional natural science-based assessment alone is neither rigorous nor rational (Stirling 2007) and needs to be complemented with social science-based 'concern appraisals' (Renn 2008).

Finally, the condition of *ignorance* is one where neither the knowledge about probabilities nor about outcomes can be made fully clear (as argued to be the case

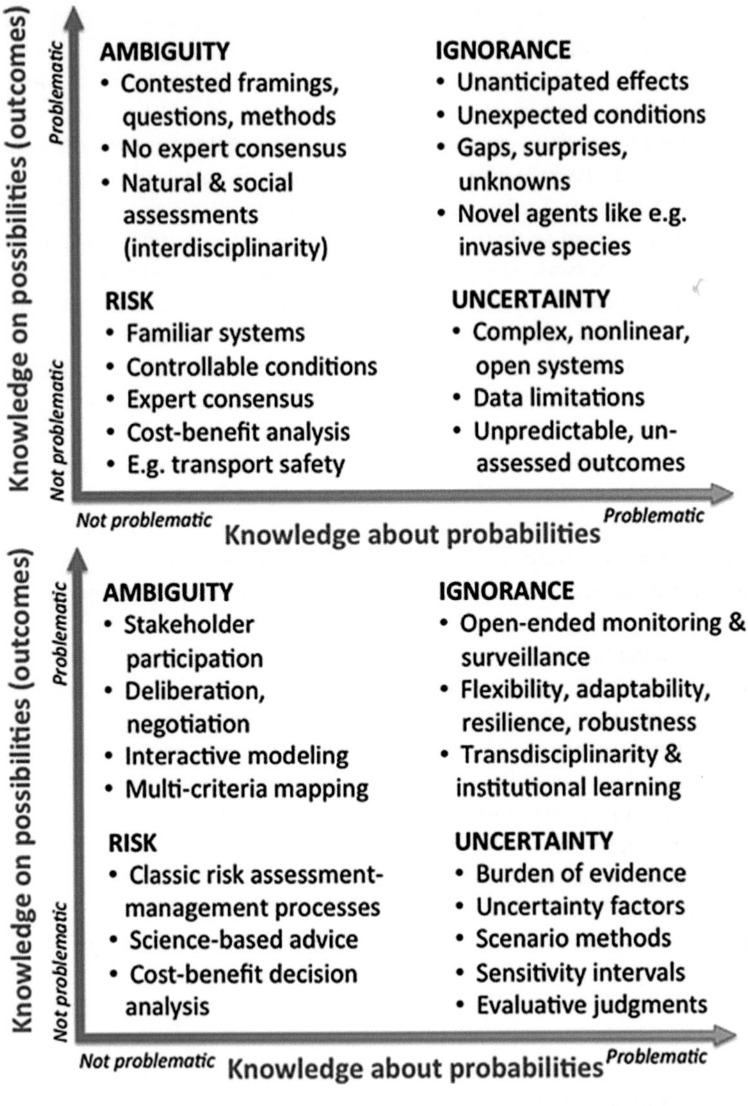

Fig. 8.2 A categorisation of four different states of incomplete knowledge (*above*) and possible responses to them for management procedures (*below*); based on Stirling (2010), see text for explanation

for invasive alien species, see Smolarz et al. 2016). Such environmental problems and risks differ from issues characterised by uncertainty in that outcome parameters cannot be pinpointed and agreed upon. Such cases also differ from environmental issues characterised by ambiguity in that knowledge about probabilities is not only contested but, often, simply unknown (Stirling 2007).

Stirling's approach to discern different types of incomplete knowledge, with regard to scientific uncertainty and sociopolitical ambiguity, has also been addressed, for example, through the concept of post-normal science (PNS) as put forward by Funtowicz and Ravetz (1993). Similar to Stirling's notion of 'ignorance', the concept of PNS proposes that under conditions of high uncertainty (limited knowledge on probabilities) and high social stakes (limited knowledge on possible outcomes), management responses to environmental problems need to seek solutions not only based on input from science, as evaluated by traditional peer review, but assembled also in an open dialogue with all affected stakeholders, something the authors call 'extended peer communities' (ibid.). These extended peer communities serve to employ additional information, knowledge and values for an active, more effective and legitimate process to identify possible solutions to specific environmental problems.

Following this classification of different forms of incomplete knowledge, Stirling also proposes a number of methodological responses that should 'illustrate the rich variety of alternatives that exist if risk assessment is not properly applicable' (Stirling 2007: 312; Fig. 8.2). However, such alternative approaches should not be taken as a 'neat one-to-one mapping of specific methods to individual states of knowledge' (ibid.) but rather serve as an array of more adequate reactions to different types of uncertainty problems that complement (and not necessarily substitute for) the traditional risk assessment-based approach.

The different categories of science-policy interactions summarised in Fig. 8.2 illustrate a more diversified picture than the 'ideal causal chain' model of applying only a traditional scientific definition of environmental problems in a linear fashion to management processes (Fig. 8.1). While traditional scientific assessment and science-based management offer powerful tools under the condition of risk (comparatively low uncertainty and known outcomes), this approach is not solely applicable to the other three categories identified by high levels of uncertainty, ambiguity and ignorance. However, contrary to such insights, the traditional scientific approach to only apply the best available (natural) scientific knowledge to policy- and decision-making often prevails and causes problems and controversy. As Stirling notes, 'persistent adherence to these reductive methods, under conditions other than the strict state of risk, are irrational, unscientific and potentially misleading' (Stirling 2007: 311).

Following this line of reasoning, we investigate whether and how different roles and approaches of science-based advice and adaptations to them in the form of management responses have evolved in our five cases of environmental governance in accordance with Stirling's typology. After analysing science-policy interfaces of the five cases in the following section, we will comparatively discuss the appropriateness of the different approaches to science and policymaking with respect to different forms of incomplete knowledge and draw conclusions on how to deal with the discovered science-policy challenges.

With respect to the theoretical context described above, our analysis focuses on two sets of questions to investigate assessment-management (science-policy) interactions in the five cases, linked to the implementation of EAM:

1. *Organisational structures* (institutional design) of the risk assessment activities and the generation, selection and implementation of management options
2. The management of *scientific uncertainties* and *stakeholder disagreements*

Under *organisational structures*, we are concerned with the *institutional* interfaces between risk assessment and management activities. We therefore analyse the regional (Baltic Sea) basis of science-policy cooperation in the five cases asking the following questions: *who is performing the assessment activities and how are assessment activities organised and carried out in committees, etc.?*

With regard to the second point, the management of *scientific uncertainties* and *stakeholder disagreements*, we investigate the science-policy interfaces with respect to how actual procedures of assessment activities are linked to the processes of giving advice and how management and decision-making bodies deal with different states of (often incomplete) knowledge and information from various sources and sectors. We also examine how stakeholder participation, knowledge inclusion and deliberation procedures are implemented.

By using this analytical framework we aim to achieve a better understanding of how governance arrangements and processes linked to various environmental problems and risks can deal effectively with the challenges of incomplete knowledge, uncertainty and stakeholder disagreements. We finally give a number of recommendations about how to possibly redesign *institutional* and *procedural structures* of science-policy interactions more effectively in the five cases of environmental governance as this will be of value also to other complex cases of environmental governance.

Methodologically, the analyses informing this chapter are based on empirical research conducted for the five case studies of the RISKGOV project,[1] which are described in the first part of this volume (Hassler 2016; Karlsson and Gilek 2016; Karlsson et al. 2016; Sellke et al. 2016; Smolarz et al. 2016). The case studies analysed the five marine environmental governance issues mentioned above with respect to three governance dimensions (see Gilek et al. 2016), one of which was 'assessment-management processes and interactions' which informed the meta-analysis presented in this chapter. For comparative purposes, the case studies used a common analytical framework and similar methodological research design (for a description, see Gilek et al. 2016; cf. Gilek et al. 2011). The results of the case studies are derived from three main data sources: document studies, semi-structured qualitative interviews (ca. 15 per case) as well as three joint thematic round-table discussions. This extensive empirical material, in combination with other research on science-policy interactions in different fields of environmental governance (see, e.g., Linke et al. 2014; Gilek et al. 2015), forms the background for the analyses presented in this chapter and allows us to draw conclusions and give recommendations for improving science-policy interactions for environmental governance of the Baltic Sea.

[1] Environmental Risk Governance of the Baltic Sea (2009–2015). See www.sh.se/riskgov

8.3 Results

8.3.1 Organisational Structures of Science-Policy Interfaces

8.3.1.1 Fisheries

Fisheries management in the Baltic Sea is primarily governed by the EU through the *Common Fisheries Policy* (CFP). The EU involves all member states around the Baltic Sea except Russia (see Sellke et al. 2016). The CFP connects the scientific assessments of fish stocks, executed on a regional basis (i.e. the Baltic Sea) by the *International Council for the Exploration of the Sea* (ICES), with management decisions taken by the EU troika of the commission, the council of ministers and the parliament. ICES works primarily for its main client, the EU Commission, and formulates requested annual scientific advice based on single fish stock assessments. Hence, regional cooperation with other management bodies such as HELCOM is rather weak and underdeveloped (Linke et al. 2014). Due to its centralised science-based structure, EU fisheries management has been depicted as 'perhaps the most top-down fisheries management system on the planet' (Degnbol and Wilson 2008: 189) and 'perhaps the most science-dependent sector in the EU' (Griffin 2009: 563). A strong dependency between science (in the form of annual fish stock assessments and subsequent fishing advice) and policy (in the form of management regulations mainly using fishing quotas) was established in modern fisheries management in the North Atlantic during the 1960s and 1970s (Gezelius 2008) and later applied to the Baltic Sea with the establishment of the CFP and an incremental EU enlargement. The institutionalised dependence between science and policy has been described as the 'TAC machine' (Nielsen and Holm 2008), which suggests that the annual single fish stock approach to management is largely incompatible with more holistic environmental governance concepts like multispecies management or EAM (Wilson 2009).

Two recent reforms of the CFP, conducted in 2002 and 2013, have aimed to change the policy system so that there is increased involvement of stakeholders in fisheries management, as well as more regionalisation on a sea-basin level, including the Baltic Sea region (Sellke et al. 2016). The 2002 CFP reform resulted in a new type of stakeholder organisation, namely, *Regional Advisory Councils* (RACs), that included fisheries representatives, NGOs and other interest groups (Linke et al. 2011). The establishment of these bodies, which give recommendations on fisheries management issues to the EU Commission, marked a new era, challenging the traditional top-down management structure of the CFP. However, while RACs have contributed significantly to a more participatory and inclusive fisheries governance in the EU, authors such as Long (2010: 294) observe that 'the impact so far of the RACs on decision-making within the CFP is less striking than their organisational structure and continues to be the subject of ongoing debate'. Overall the establishment of RACs can be seen as a substantial shift towards stakeholder involvement and potentially even a partially delegated management responsibility to the industry or other actors - aspects that have been more vividly discussed in the context of the new CFP reform in 2012/13 (Hatchard and Gray 2014; Nielsen et al. 2015). However, both the RACs' organisational design within the current CFP structure as well as the interactive processes occurring (e.g. in the Baltic RAC) heavily impact on their role and function in future CFP systems. They still have to show how they can deliver future progress and consolidate their role as responsible actors in the transformation of fisheries management towards new approaches of co-management characterised by a new burden of proof regime and sharing of power and responsibility (Linke and Bruckmeier 2015; Linke and Jentoft 2013, 2014).

8.3.1.2 Eutrophication

Eutrophication management at the regional Baltic Sea level is also highly, if not exclusively, science based, however not via an EU centralised policy as for fisheries but through a more informal regional institutional arrangement. The interplay between science and policymaking has been established via a specifically developed decision-support system called 'NEST' with HELCOM as the responsible management agency (Wulff et al. 2007) and with links to various EU directives (see below). As described in more detail by Karlsson et al. (2016) and Linke et al. (2014), scientists at the Baltic Nest Institute developed NEST as a model to provide scientific recommendations for transnational eutrophication management in the Baltic Sea in close cooperation with HELCOM (cf. Wulff et al. 2007), an interplay referred to as the 'NEST-HELCOM nexus' (Linke et al. 2014). This tight science-policy interface, realised through the close interplay between the Baltic Nest Institute and HELCOM, was inscribed in the eutrophication segment of the Baltic Sea Action Plan (BSAP) (HELCOM 2007; Linke et al. 2014) and has been described as a great success in terms of science-policy interaction because the promotion of scientific research resulted in improved knowledge that was, through the decision-support system NEST, made directly applicable to policy- and decision-making on how to mitigate the harmful effects of eutrophication in the Baltic Sea (Johansson et al. 2007). As opposed to the EU fisheries policy (CFP), EU policies applied for mitigating eutrophication (e.g. the Nitrates Directive and the MSFD Descriptor 5 on eutrophication) rest primarily with the member states as does the implementation of agreements and recommendations under regional sea conventions such as the Helsinki Convention. Regarding the latter, the perceived consequences of potential nationally initiated measures led to serious criticism, for example, from farmer organisations (BFFE 2013; LRF 2013), which distorted the previously successful and exclusive interplay between science and policymaking in this area. However, within the EU arena, the MSFD is still under implementation and it remains to be seen if policies fully based on scientific findings will be implemented in full, in order to promote the far from achieved targets for mitigating eutrophication (Karlsson et al. 2016; HELCOM 2013). Similarly, the BSAP-related scientific assessments are linked to various EU directives[2] for a possible harmonisation of aims and objectives (Andersen et al. 2011), but the ultimate success of eutrophication management strongly rests with *national implementation plans*. It is yet unclear how these will be further implemented (Gilek et al. 2013; Pihlajamäki and Tynkkynen 2011).

The inclusion of stakeholders into processes of management and policymaking with regard to eutrophication in the Baltic Sea is still underdeveloped. While HELCOM has moved from an observer strategy to stakeholder dialogue forums, farmers both nationally and internationally have raised their concerns in connection with the most recent HELCOM ministerial declaration in 2013. Farmers want to be more involved and voice their opinions about ecological and scientific as well as socio-economic aspects related to the implementation of the BSAP (BFFE 2013; LRF 2013; see Sect. 8.3.2).

8.3.1.3 Invasive Alien Species

Invasive alien species (IAS) are recognised as one of the most severe threats to marine biodiversity worldwide. Still, management of environmental risks connected with IAS is a new and undeveloped field of environmental governance (cf. Smolarz et al. 2016). Shipping, which primarily via ballast water has been the source of approximately 50 % of nonindigenous species found in the Baltic Sea (Leppäkoski and Laine 2009), is currently the prime focus in attempts to develop regulations to manage IAS risks in the marine environment. At the core of these regulatory developments is the international BWM Convention,[3] which sets up standards and procedures for the management and control of ships' ballast waters and sediments to prevent the spread of harmful aquatic organisms. However, although the BWM Convention was opened for signatures by the International Maritime Organization (IMO) in 2004, it is not expected to enter into force until 2015–2016 at the earliest due to a slow ratification process in many countries.[4] Still, substantial efforts have at the same time been made by intergovernmental organisations such as the EU and HELCOM to improve regulations on IAS, as well as to facilitate implementation of the BWM Convention, for example, by providing guidance and science support adapted to the European and Baltic Sea contexts. The EU has recently adopted a regulation on IAS (EU 2014) based on prevention, early warning and rapid response and management, leaving it to member states to establish science-based lists of IAS. Moreover, HELCOM (2014) has drawn up voluntary recommendations for

---

[2] Examples of such EU directives are the Water Framework Directive (WFD), the Marine Strategy Framework Directive (MSFD), the Nitrates Directive and the Urban Wastewater Directive.

[3] International Convention for the Control and Management of Ships' Ballast Water and Sediments; see http://www.imo.org/OurWork/Environment/BallastWaterManagement/Pages/BWMConvention.aspx

[4] The convention enters into force 12 months after ratification by 30 states, representing 35 % of world merchant shipping tonnage; at the time of writing, in May 2015, 44 states with 32.9 % of the tonnage have done so.

how to safely perform ballast water exchange before entering the Baltic Sea, as well as provide guidance on how to perform risk assessments and monitoring. Various research projects and databases have been initiated at the European and Baltic Sea level to compile inventories of IAS, as well as identify species and shipping practices connected with particularly high levels of environmental risks (HELCOM 2014; Smolarz et al. 2016).

Hence, both management and assessment activities linked to IAS risks in the Baltic Sea are, despite some recent progress, still in need of further development, particularly in order to reach the ambitious objective set by HELCOM in the 2007 Baltic Sea Action Plan of no introduction of IAS from ships. For example, observations of new IAS often still build on incidental reporting, since targeted and coordinated monitoring is often lacking in high-risk areas. In a wider perspective, it has also been argued that management and science support needs to be improved for 'non-shipping' sources of IAS such as aquaculture and that interdependence with other environmental issues such as fisheries, human-induced climate change and eutrophication need to be considered in risk assessments (Smolarz et al. 2016).

Wide stakeholder involvement at the science-policy interface is lacking in the case of IAS. For most stakeholders and the general public, this seems to be due to a widespread lack of interest in the IAS issue and its management, which are seen as rather uncontroversial and straightforward. In contrast, stakeholders associated with the shipping and cargo sector have shown a strong interest in contributing knowledge and opinions as part of negotiations on development and implementation of regulations (Lemke et al. 2010). Negotiations leading up to the 2004 opening of the BWM Convention have been described by IMO as 'complex'. The subsequent rather long ratification process seems to have been influenced by discussions on, for example, technical possibilities and costs of ballast water management. It therefore remains to be seen whether or not these sector-based discussions will improve efficiency and effectiveness of IAS-related management in the Baltic Sea.

8.3.1.4 Chemical Pollution

There are two central spheres for the governance of chemicals in the marine environment within Europe of relevance for the Baltic Sea region: chemicals policy that takes a market perspective and environmental policy that is based on, for example, aquatic parameters which aim to protect health and the environment (Karlsson et al. 2011).

Within the market sphere of chemicals policy, substances have been classified and further regulated in the EU since the 1960s. Over time, a strong risk-based foundation has emerged, implying that unacceptable risks must be proven in comprehensive assessment before risk reduction measures could be motivated. The ultimate manifestation of such a view is the *Technical Guidance Document* (TGD) of the EU, comprising thousands of pages of instructions for scientific risk assessments (see, e.g., ECB 2003), which is applied by experts in various advisory and decision-making bodies. The heavy burden of proof associated with this 'TGD machine' caused a science-policy deadlock in the 1990s, resulting in assessment of less than 100 out of 100,000 substances registered on the EU market (see Karlsson 2006). After lengthy political debates, the EU adopted a new policy for the bulk of industrial substances, namely, the REACH regulation, which is now the central piece of chemical law in the Baltic Sea region today (EC 2006). REACH is EU harmonised and charges industry with the responsibility of registering data on substance properties. The data is to be evaluated on a scientific basis by the European Chemicals Agency (ECHA), which has a strong technical and expert orientation, and competent authorities of individual EU member states, which judge whether or not to suggest risk reduction measures such as authorisation or restrictions. A strong burden of proof still rests with regulators and, consequently, the implementation has been slow (Karlsson 2010). Several other laws, besides REACH, regulate substances in general and more specifically chemical substances in specific products, for example, electric and electronic products (e.g. EC 2003).

The second sphere, environment-oriented policy and law, focuses on specific parameters for, e.g., water quality, as stated in the WFD (EC 2000) and the MSFD (EC 2008). The scientific focus is strong in this sphere as well, but the starting point here is health and the environment rather than market aspects and EAM is, therefore, often applied. Risk assessments – or more commonly, environmental assessments – are carried out in a number of settings, e.g., by agencies, universities or international bodies such as HELCOM, the latter also adopting recommendations on, for example, restrictions aimed at parties of the convention. Individual countries, within the general framework set up in law, are then commonly expected (in the case of HELCOM) or charged (in the case of EU) to ensure implementation of different risk management measures, for instance, regulating emissions of substances from various sources.

There are no specific organisations of a participative nature with regard to stakeholders pertaining to the environment or chemical industry. However, some representatives of certain stakeholder groups are invited to and involved in various steps in decision-making procedures regarding both assessment and management issues, more frequently the case in public organisations at the national, EU and international level than in the scientific committees under them. The strong burden of proof that is placed in the public domain has given industry stakeholders a favourable position to delay processes by repeatedly demanding more data.

8.3.1.5 Oil Discharges Linked to Marine Transportation

Oil transportation in the Baltic Sea creates two different kinds of environmental risks, namely, accidental and intentional oil spills (Hassler 2011, 2016). Whereas the former are rare but may have severe negative impacts on local or regional ecological systems or result in major economic loss and social disturbances, the latter consist of the many small acts of pollution that result from operators cleaning tanks or flushing machine compartments *en voyage* without taking proper care of the disposal of oily residuals. Although the governance structures and measures for oil spill control, as well as associated science-policy interactions, differ in many respects with regard to these issue areas, it is clear that marine transportation is generally governed by a relatively straight chain of command via global conventions under the UN agency IMO (Hassler 2016). However, implementation of these global conventions is primarily the responsibility of individual countries through flag and port state control. Hence, in the Baltic Sea, the management of oil spill risks is determined not only by global conventions but also to a large extent by the individual coastal states, as well as intergovernmental organisations such as HELCOM and the EU. Such management measures at the Baltic Sea level include port state control of ship safety and an incentive-based 'no-special-fee' system to promote safe waste delivery and tank cleaning at port,[5] as well as aerial surveillance of oil spills (Hassler 2016).

The case of oil discharges involves a comparably clear separation of assessment and management activities, where assessment primarily takes the form of monitoring and surveillance of vessel functionality, tanker traffic and illegal discharges. While these regional, subregional and unilateral monitoring activities[6] appear to be more developed than e.g. in the case of IAS, coordination still constitutes a significant challenge. This is, for example, seen in the uneven number of flight hours reported by different countries in aerial surveillance of oil discharges (Hassler 2016). The assessment and monitoring linked to tanker traffic and oil discharges do not, however, appear to be very influential in management discussions and decisions linked to technical safety requirements such as double hulls, separate ballast tanks, better navigation equipment, etc. These discussions are instead mostly carried out in IMO, the EU, member countries and international shipping organisations such as INTERTANKO (Hassler et al. 2010).

Therefore, stakeholder involvement in shipping is considerable at the international level, where sector organisations both provide important knowledge and try to influence regime outcomes in the directions they prefer. Classification societies and insurance companies also play important roles in the modernisation of oil tanker fleets. However, stakeholder influence and participation, as well as other forms of civil society involvement, are considerably less intensive at lower governance levels in the Baltic Sea region.

[5] This is 'a charging system where the cost of reception, handling and disposal of ship-generated wastes, originating from the normal operation of the ship… is included in the harbour fee or otherwise charged to the ship irrespective of whether wastes are delivered or not' (HELCOM recommendation 28E/10, paragraph 1.1).

[6] Examples of monitoring activities linked to oil discharges and tanker safety in the Baltic Sea include the HELCOM AIS (real-time automatic tracking of all larger vessels), air surveillance and hydrographical surveys (Hassler et al. 2010).

8.3.2 Managing Uncertainties and Stakeholder Disagreements

8.3.2.1 Fisheries

In EU fisheries governance, disputes and conflicts exist between major stakeholder groups such as fishermen, NGOs, scientists and policymakers about how to manage and/or preserve fish stocks adequately in the Baltic Sea. These conflicts revolve around basic issues such as 'whose knowledge counts' in the debate between conservationists and those who represent fisheries' interests. Stakeholders are often dissatisfied for different reasons with the management process and consider separate aspects of the CFP as the cause for the failure to attain more sustainable fisheries practices in the Baltic Sea. The disagreements can be categorised under two major issues: the accuracy, objectivity and reliability of different types of knowledge about fish stocks (i.e. data uncertainties), and therefore their applicability to decision-making, and secondly, how social issues, i.e. the different value perspectives and worldviews of stakeholders and socio-economic and cultural dimensions, shall be addressed in policy and management under the CFP, which has traditionally been exclusively natural science based (Linke and Jentoft 2014).

Uncertainties in fish stock assessments, and associated stakeholder disagreements about how these uncertainties should be treated in management, have been categorised into three main sources (Linke et al. 2014). The first is the classic notion of scientific uncertainty relating to lack of data, natural variability and ecosystem complexity that is creating tensions in the scientific community (e.g. within ICES) about adequacy of data sources and how to present uncertainty while giving advice to management bodies like the EU Commission (Wilson 2009, 123ff). A second site of uncertainty exists at the centre of science-policy interactions: whereas managers and decision-makers usually want clear, quantified advice, for example, in the form of a fishing quota, scientists often like to give more nuanced qualitative assessments, referring, for example, to 'poorly understood stock dynamics', 'problems with estimating discards' or 'changing fishing patterns' (Sellke et al. 2016; Wilson 2009: 125). This issue relates to the basic conceptual problem of how to define the roles of science and policy in practical interactions (see theoretical Sect. 8.2 above) and leads to constant negotiations about how and where to draw the science-policy boundary – both within science (i.e. in ICES) and management, as well as in the wider stakeholder arena. The third aspect of uncertainty relates to problems that emerge when stakeholders interpret scientific assessments differently according to their own interests and blame opposing actors of misunderstanding or misinterpreting science. Such varying interpretations of scientific uncertainty are obvious in negotiations, e.g., within the RACs, where fisheries and NGO representatives often strongly disagree with each other about whether and how the lack of data and insufficient knowledge shall be used to restrict fishing activity according to the precautionary approach of the CFP, to which conservation groups usually adhere (Linke and Jentoft 2013, 2014).

These three sites of uncertainty and stakeholder disagreement in EU and Baltic fisheries management reveal key challenges of dealing with highly politicised cases of environmental governance inflicted by social and economic values within a classically designed science-policy interface.

8.3.2.2 Eutrophication

Eutrophication and related management objectives to combat it comprise enormous complexities not only in ecosystem functioning but also with respect to a diverse political arena of governments, economic sectors and stakeholder groups affected by strategies to control eutrophication via nutrient reductions. There are ecosystem structures that lead to particular environmental phenomena in the Baltic Sea such as a relatively slow turnover rate of water (approximately 30 years) and nutrient storage and release from sediments. Furthermore, there are indications that large-scale environmental change (so-called regime shifts) can be amplified by, for example, overfishing of cod (Casini et al. 2009). These complexities lead to several complications for science-policy interactions such as long time lags between implementation of management measures and potential positive effects on environmental status. Such time lags open up for a critique of the scientific basis of taken and proposed measures as well as with regard to general complaints over policies for nutrient reductions, since it is uncertain if and when the desired environmental objectives can be reached. Despite the complexity of the natural and social systems linked to eutrophication management in the Baltic Sea, interactions between scientists and policymakers were quite smooth for a period of time, even if serious questions are now being raised by stakeholders (see below). While a controversy existed earlier (mostly between scientists) about which nutrients needed to be primarily controlled to combat eutrophication – nitrogen, phosphorus or both – there is now a general consensus that both nutrients need to be controlled (Conley et al. 2009a).

However, particularly at the national implementation level, discussions are ongoing about which objectives to set and which nutrient reduction measures might be most cost effective (e.g. Elofsson 2010; Gren 2008; LRF 2013). Moreover, so-called engineering approaches to counter eutrophication problems, such as chemical sequestration of phosphorus or artificial oxygenation, are put forward by various actors but are also criticised by most marine scientists as inappropriate to combat large-scale offshore eutrophication in the Baltic Sea (Conley et al. 2009b; Conley 2012). The academic debate regarding whether man-made oxygenation may or may not be a means to reduce eutrophication effects in the Baltic Sea is still not over (cf. Stigebrandt and Kalén 2013; Stigebrandt et al. 2014). Although substantial uncertainties exist regarding eutrophication assessments and advice, as well as about the ways in which management should best distribute the costs of nutrient reductions among Baltic countries and stakeholders within them, lack of data, scientific uncertainty and stakeholder disagreements have, in contrast to fisheries, not yet sparked similar levels of conflict and public controversy about appropriate policies, management objectives and political decisions.

This situation, with comparably low levels of stakeholder disagreement on eutrophication management during the first part of the BSAP process leading to a decision on country-wise nutrient reduction targets, might be due to largely underdeveloped forums for stakeholder participation and representation in eutrophication-related policy and management. However, affected stakeholder groups like farmers are increasingly organising themselves to voice their opinions, concerns and recommendations on nutrient reduction strategies, at a national as well as international level. These farmers' (as one of the most affected economic stakeholder groups) viewpoints involve calls for inclusion of socio-economic assessments in the implementation of the BSAP, as well as initiation of a broader dialogue with society 'that should be designed so that real influence could be exercised without having to take part in numerous scientific and other meetings' (BFFE 2013).

With regard to the initial assessment for the BSAP in 2007 involving the tight and exclusive NEST-HELCOM interaction (see above and Karlsson et al. 2016), Swedish farmers voiced concerns only with respect to the fact that a 'new and previously untested model is used as the basis for a multi-million decision' (LRF 2010). Now the same stakeholder group, as well as its international counterpart, the Baltic Farmers' Forum on Environment (BFFE), criticises current eutrophication management strategies for implementing BSAP in a more forceful and detailed way with respect to (*1*) the neglect of the internal phosphorus loads from sediments in the Baltic Sea for BSAP, (*2*) unrealistic estimations of time frames for recovery (cf. Gustafsson et al. 2012) and (*3*) cost-efficiency issues (BFFE 2013; LRF 2013).

8.3.2.3 Invasive Alien Species

Despite the fact that maritime transportation is nowadays recognised as the main vector of marine IAS (Leppäkoski and Laine 2009; Smolarz et al. 2016), there are numerous uncertainties associated with the identification of other potential pathways, high-risk species and, above all, long-term ecosystem impacts and associated socio-economic consequences. Linked to this is also an interpretative ambiguity related to assessing the consequences of IAS, since a particular introduction of, for example, a fish species could be associated with both negative (e.g. on native species) and positive effects (e.g. on commercial fisheries). Moreover, it has been argued that neither knowledge about probabilities nor about environmental impacts can be made fully clear for IAS (i.e. in line with an *ignorance* type of risk, cf. Sect. 8.2), since ecosystem outcomes of IAS will always be influenced by context-specific factors such as type and condition of species, location and season of introduction, etc., associated with particular species introduction (e.g. Zavaleta et al. 2001).

Even though IAS is characterised by high levels of scientific uncertainty and interpretative ambiguity, these aspects have not provoked the same level of disagreement and controversy in science and management as, for example, in fisheries (e.g. Lemke et al. 2010; Linke et al. 2014). This does not mean that there have been no disagreements or controversies linked to regulatory developments in the IAS case but rather that such disagreements have had causes other than scientific uncertainties about ecosystem impacts and consequences. For example, disagreements on the technical and economic feasibility of various suggested options for ballast water management have been common (cf. Lemke et al. 2010).

In their analysis of IAS management and assessment, Smolarz et al. (2016) conclude that the fundamental uncertainties outlined above are not sufficiently acknowledged, assessed and managed in the Baltic Sea. The authors also point to the general need to substantially improve the knowledge base, as well as procedures for implementing a precautionary and ecosystem-based approach to IAS management. Still, despite these significant challenges, it is possible to observe some initial steps that have been taken to manage ambiguities and uncertainties linked to IAS risks in the Baltic Sea.

First, by commonly defining IAS as '…species whose introduction and spread threatens ecosystems, habitats or species with economic or environmental harm', the interpretative ambiguities seem to have been 'defined away' in what can be seen as a pragmatic and precautionary approach (cf. Smolarz et al. 2016). That is, this definition assumes that all IAS per definition could cause harm.

Second, although uncertainty in the scientific discourse is mainly discussed in terms of a substantial lack of data (e.g. on monitoring of nonindigenous species in particular environments and ecological consequences of these), the development of regulatory frameworks focuses on preventing entry of IAS in the Baltic Sea by requiring, for example, shipowners to manage ballast water safely. It is, however, too early to evaluate the success of these chosen management approaches, since safety regulations are still not fully implemented and several exemptions have been discussed, for example, in HELCOM regarding low-risk shipping routes in the Baltic Sea.

Finally, several scientific projects and risk assessments initiated by, for example, HELCOM and EU research programmes, have addressed uncertainty by attempting to develop strategies for screening and prioritising risks of various species, vectors, areas and routes. HELCOM has since 2008 published a list of harmful species. There is also a prioritised list of 'target' species exhibiting properties leading to high environmental risks (e.g. HELCOM 2014). Such a pragmatic prioritisation strategy clearly has the potential to focus efforts and resources on issues and areas exhibiting high risk. However, concerns have also been raised that an overly strict management focus on known risks may counteract precaution (Smolarz et al. 2016).

Hence, although uncertainties linked to IAS risks and their consequences for management in the Baltic Sea are substantial and seldom fully acknowledged, some rudimentary pragmatic steps for addressing uncertainty have recently been taken. It is of course too early to tell whether or not these approaches for uncertainty management will be sufficient to counteract potential disputes among stakeholders and allow for effective implementation of the global BWM Convention and the recently adopted EU regulation on IAS in the Baltic Sea region. Ultimately, it also remains to be seen if the primary focus on IAS risks connected with shipping is sufficient to reduce the overall environmental risks of introducing IAS from all sources to levels enabling publicly decided targets.

8.3.2.4 Chemical Pollution

The complex web of policies, laws and stakeholder groups with different ambitions, in combination with deep uncertainties about both exposure conditions and the inherent properties of thousands of substances on the market, results in possible controversies with regard to chemical pollution. Only few substances on the market have been thoroughly assessed with regard to health and environmental risks (Allanou et al. 1999; Gilbert 2011). Moreover, risk assessments are routinely done substance by substance, according to the highly structured TGD (see above), overlooking how substances may interact despite evidence of special effects of exposure to mixtures ('cocktail effects'; see Kortenkamp et al. 2009). All in all, this makes chemical policy more about managing uncertainty than about managing well-studied risks. Lack of knowledge results in that different actors are competing over preferential rights to interpret the incertitude, which consequently leads to politicisation and disagreements, even within so-called impartial expert groups (Eriksson et al. 2010a, b). Such controversies were obvious not least during the setting up of the REACH regulation in the 1990s and 2000s. While the regulation was considered one of the most contested pieces of legislation in the history of the EU (Fisher 2008; Selin 2007), still today, no specific forum exists for dealing with stakeholder disagreements, at least not beyond conventional representation, consultations and meetings.

In the market-oriented sphere of chemicals policy, stakeholder disagreements and conflicts arise along several lines, for example, between proactive and reactive member states and EU institutions and between other stakeholders such as NGOs and business organisations (Eriksson et al. 2010a). The debates centre on the health or environmental impact of a specific substance and what principles should guide decisions under uncertainty. NGOs and some environmentally ambitious member states support the precautionary principle, i.e. to err on the side of safety and allow for basing measures on intrinsic hazardous properties, whereas industrial organisations commonly promote a non-precautionary, solely risk-based approach, in which the burden of proof is placed on the regulator. This means that management measures must be based on proven unacceptable risks. In contrast to fisheries, where opposition to the dominant set-up of science policy comes from a user perspective (fishers' organisations), the science-policy interface in chemicals is criticised more from an environmental point of view.

In the more recent and environment-oriented policy sphere (mentioned in Sect. 8.3.1.4, including MSFD and BSAP), the precautionary principle is often explicitly recognised. Hence, the mere presence of a substance, suspected to be problematic for the environment, commonly motivates requests for preventive measures. This comparatively low burden of proof for public policymakers implies a very different science-policy interface than in the sphere of chemicals policy. However, nowadays some precautionary elements can be found also in the REACH regulation, which allows for decisions on preventive measures, such as authorisation of substances that are toxic or very bioaccumulative and persistent, even if there is uncertainty about exposure conditions. Such precautionary measures in cases of scientific uncertainty are the exception rather than the rule in chemicals policy, and traditional requirements of establishing comprehensive risk assessments before restrictions can be decided are strong (cf. Karlsson and Gilek 2016).

In summary, there are clear differences in how uncertainty and disagreements are dealt with (precautionary or not) in the environmental and the market spheres of chemicals policy, respectively. Science-policy interfaces are hence still far from mature and well coordinated and the deep uncertainty in this governance domain is not addressed rationally. In addition, there are no (standing) forums where stakeholders regularly can meet and discuss these issues and how best to handle disagreements.

8.3.2.5 Oil Discharges Linked to Marine Transportation

The short- and long-term ecosystem effects of oil discharges depend on a variety of factors including the season and weather, type and quantity of oil spilled, habitat, type of shoreline as well as the tidal energy and type of waves in the area of the spill (e.g. Rousi and Kankaanpää 2012). Consequently, there is a substantial degree of scientific uncertainty linked to the assessment of ecosystem risks of oil discharges. However, this type of scientific uncertainty on ecosystem impacts is not of substantial importance for assessment-management interactions in this case (Hassler et al. 2010). That oil discharges are dangerous for the marine environment and should be avoided as much as possible within reasonable economic limits is not contested.[7] The main issues of assessment-management interactions relating to oil spills rather concern (*1*) what the actual probabilities and prevalence of oil discharges are (i.e. mainly monitoring and surveillance activities) and (*2*) how the probability, extent and prevalence of oil spills can be reduced through improved technological safety, reductions in human errors and improved management measures to reduce intentional oil discharges. Consequently, compared with risks such as chemical pollution, the risks of oil discharge and uncertainty-related challenges concerning them are not as severe in the context of the Baltic Sea, barring a few exceptions.

First, there is still a lack of aerial and satellite data of intentional oil discharges for monitoring and surveillance purposes in spite of a well-developed system for monitoring of larger vessels through the HELCOM AIS system (Hassler et al. 2010). This uncertainty linked to surveillance and monitoring impedes a comprehensive assessment of the prevalence of oil discharges (especially intentional discharges), as well as complicates attempts to enforce and monitor the efficiency of management strategies (such as the above-mentioned HELCOM no-special-fee system) (Hassler et al. 2010; Hassler 2011).

Second, regarding safety improvements, there is scientific uncertainty linked to issues concerning the feasibility and cost-efficiency of various technical solutions.

[7] It can also be argued that the PSSA (particularly sensitive sea area) classification of the Baltic Sea by the IMO shows that there is an agreement (and substantial scientific knowledge) that the Baltic Sea is particularly sensitive to, e.g., oil discharges.

Today, discussions on these management issues appear to mainly take place in rather closed settings, among IMO officials, IMO member countries and representatives from the shipping industry with limited participation from environmental NGOs (Hassler et al. 2010; Hassler 2016).

Finally, the ecological sensitivity of different areas and 'marine crossroads' where incidents are more likely to occur can be comparably easily identified and located. It is, however, more complex to relate these parameters to what we know about the likelihood of human errors. Although clear distinctions between human error on the one hand and technical malfunction on the other can seldom be made, available statistics show that the former tend to still be the most common cause of incidents and accidents (Knudsen and Hassler 2011).

8.4 Summarising Discussion

Using the 'uncertainty matrix' presented in Sect. 8.2 with its four idealised states of incomplete knowledge (Fig. 8.2), we can categorise our five cases as belonging to the categories of risk, uncertainty, ambiguity or ignorance (Table 8.1).

Our analysis focused on two main aspects of science-policy interfaces in the five cases: (*1*) the *organisational structures* of science-policy interfaces and (*2*) the management of *scientific uncertainties and stakeholder disagreements*. This investigation revealed substantial differences in terms of institutional design of assessment-management interactions, as well as in terms of how scientific uncertainty and sociopolitical ambiguity, stakeholder conflicts and controversies are addressed. Whereas, for example, chemical risks are associated with paramount uncertainty, oil discharges are not. Fisheries management on the other hand involves a high degree of sociopolitical ambiguity, whereas in the case of IAS the opposite is true. Finally, oil transportation fits with the more traditional ('technical') risk type. Our analysis also reveals a great deal of variation in societal responses to the cases and that these responses are often motivated by factors other than the actual risk characteristics.

Regarding the *organisational structures* of science-policy interfaces, we can conclude that different forms of institutions and institutional arrangements have evolved over time. We find these structures relatively well-formalised in the fisheries case (via the EU's CFP system); rather informal and bilaterally developed for combating eutrophication (through the interaction of the 'NEST-HELCOM nexus'); largely underdeveloped in the IAS case, split into different spheres in the chemicals area (environment versus market); and seemingly straightforward in the case of oil transportation (through country surveillance and monitoring). We also see a clear trend in terms of an intensified role of EU cooperation over time as countries are bound to a centralised EU policy in fisheries management (CFP) and various EU directives and strategies in the other cases. However, the role of the EU varies in the five issues. Furthermore, we can identify different forms of dependence on experts across the cases. In fisheries management, a highly institutionalised formal linkage exists between the science advice system (ICES) and the EU Commission

Table 8.1 Summary of observed assessment-management interactions linked to environmental problems and risks in the Baltic Sea

Environmental risk case	State of incomplete knowledge	Organisational structures of assessment-management interactions	Management of uncertainty and stakeholder (SH) disagreements
Fisheries	Ambiguity/Uncertainty	Highly formalised structures linked to the EU's CFP; science-based with formalised stakeholder consultation e.g. via RACs	Different framings and perceptions of risk (environmental vs. socio-economic); controversy among SH on dealing with uncertainty (e.g. in data, assessment and interpretation)
Eutrophication	Uncertainty/Ambiguity	Strong/exclusive role of science advice (co-production via NEST-HELCOM); limited stakeholder involvement	Science-based assessment; diffuse or instrumental treatment of uncertainty; increasing disagreements among SH groups
Invasive alien species	Ignorance	Rather undeveloped; no clearly identifiable stakeholder groups	Science-based risk assessment; undeveloped uncertainty treatment; no major controversy
Chemical pollution	Uncertainty/Ambiguity	Strict formal demands on risk assessments by public bodies in chemicals policy, less so in environmental policy; limited stakeholder involvement	Science-based assessment of risks in some contexts, precautionary approaches in other; controversy on how to cope with deep-rooted uncertainty; SH conflicts but not coordinated
Oil discharges	Risk	Focus on surveillance and monitoring (IMO flag & port state control)	Science-based assessment of risks; focus on technical safety analysis, control & enforcement

(policymaking), as a result of a historic path dependency, that in turn causes various forms of institutional inertia (Hegland and Raakjær 2008; Wilson 2009), even though recent developments with the RACs such as the Baltic RAC may gradually change the picture. This path dependency and the resulting inertia of science and policymaking can also be seen in the chemicals case, where the traditional primacy of risk-based approaches (e.g. the 'TDG machine') is strong. This phenomenon cannot be observed to an equal degree either in the case of eutrophication, where the NEST-HELCOM model offered a comparatively new approach of science-policy interaction, or in the emerging domain of IAS, where an institutionalised form of assessment-management interaction is not yet in place.

Furthermore, well-developed institutional structures as they presently exist for EU fisheries management also seem to allow for the establishment of improved deliberative processes, which at least potentially could result in improved analysis of the socio-economic dimensions of the respective domain of marine governance (cf. Linke and Jentoft 2014; Urquhart et al. 2014). However, apart from an incremental 'opening up' of the linear science-policy structures in fisheries, we cannot find such tendencies developed to a similar extent for eutrophication, IAS, chemicals or oil transportation. In the cases of eutrophication, IAS and chemicals, it may also be substantially harder to identify clear groups of actors definable as 'legitimate stakeholders' to be included in the policy system than when one particular group of actors, for example, 'resource users' (fishermen) or oil transporters (shipowners), can be defined as such and often accused of causing problems e.g. by the media. A rather clear identification of legitimate stakeholders, as we see in fisheries management, is not apparent in the cases of eutrophication (where farming only represents one of several important nutrient sources), chemicals (with a complex web of chemical producing and using actors), IAS or oil transportation.

Regarding *uncertainties and stakeholder disagreements*, the observed assessment-management interactions do not seem to adequately address the key challenges posed by the different states of knowledge as described in Fig. 8.2 and the sociopolitical ambiguities involved. With regard to management responses dealing with different forms of uncertainty and stakeholder conflicts, thorough analyses of the specific risk characteristics are not apparent, i.e. of the social and natural knowledge requirements or of the suitability of different management strategies such as a traditional science-based assessment, a precautionary approach, deliberative methods for stakeholder participation or extended peer review, etc. (see Sect. 8.2 and Fig. 8.2). Instead, all the studied assessment-management interactions can be classified as building primarily on traditional *science-based assessments*, which are then applied to policy- and decision-making (Table 8.1). We see, however, a number of ongoing changes and developments in each of the cases that illustrate adaptive responses to the key management challenges studied here.

For example, we detect advancing processes of deliberation and stakeholder inclusion in both the science and management sector in the fisheries case, as an attempt to deal with the principal challenges posed by uncertainty in assessments and management, as well as to address conflicts of interests among the key actors. Concerning eutrophication, we can see similar challenges emerging (i.e. stakeholder conflicts impacting on current science-policy interactions). However, in the eutrophication case, stakeholder groups are not yet organised to a similar extent as in fisheries management, and this case does not experience protests and debate to a similar degree as the (much older) governance domain of fisheries. In the chemicals case, where conflicts are dispersed among a variety of stakeholders, we find some approaches for applying precaution to handle the paramount uncertainties and ambiguities. With environmental governance linked to oil transportation, we see a case of technical risk management focusing on safety regulations and reducing human errors, including surveillance and monitoring, without any stakeholder conflicts between dominant actors. IAS governance was, contrary to most of the other cases, identified as an area of low public interest and minor stakeholder conflicts. An interesting observation is that EAM was commonly invoked by stakeholders in the IAS case (albeit often with diverging definitions and framings) as an appropriate basis for developing assessments as well as management measures. Such an interest in EAM was apparent in many of the other cases as well. However, while attempts to develop EAM are perhaps most visible in the environment-oriented policies of the chemicals case and in the struggle to cope with interactions between eutrophication on the one hand and fisheries and chemicals on the other, practical conclusions on how to effectively address and integrate EAM in concrete assessment-management interactions (science-policy interfaces) are still largely missing.

Juxtaposing our results to the model of incomplete states of knowledge in the 'uncertainty matrix' presented above and proposed methodological responses (Sect. 8.2), we cannot yet find a variety of alternative approaches that according to Stirling are required when the traditional risk assessment-management approach is inappropriate. Nevertheless, the comparative analysis of our five marine governance cases highlights different stages in the adaptation of such methodological responses to uncertainty, ambiguity and ignorance that may have the potential to 'reveal the intrinsically normative and contestable basis for decisions, and the different ways in which our knowledge is incomplete' (Stirling 2007: 312). Such more reflective perspectives on the role of science and/versus policy- and decision-making are important for developments to incorporate all three pillars of EAM – ecological, social and economic sustainability – in the respective management systems.

To summarise, our study shows that assessment-management interactions in general have neither developed in line with theoretical assumptions nor in an always purposeful manner. However, a continuation of the tendencies, trends or advancements discussed here can lead to increasingly coordinated and more effective risk governance policies in all five cases. In addition, the ongoing Europeanisation of marine environmental governance in the Baltic Sea illustrated in other studies (cf. Gilek et al. 2015) could potentially lead to a diminished role for HELCOM, owing to the EU membership of most countries around the Baltic Sea and stronger legislative powers of the EU. This will potentially complement and strengthen the mentioned trends. At the same time, the ongoing regionalisation of environmental governance of the Baltic Sea, as seen, for example, in the EU strategy for the Baltic Sea region or with BS RAC under CFP and more recently also the member states' forum BALTFISH (cf. Sellke et al. 2016), may also strengthen the role of HELCOM. HELCOM's BSAP is therefore seen as an instrumental way to implement, for example, the EU's MSFD (Gilek et al. 2015). If so, both the EU and HELCOM could become more important for Baltic Sea environmental governance in the future. This might, however, not be an identical outcome for all cases. It remains to be seen how these diverse, but potentially mutually supportive, processes will influence science-policy interfaces linked to the governance of the Baltic Sea environment. Some conclusions and recommendations from the present study are nonetheless highlighted in the following section.

8.5 Conclusions and Recommendations

Our study shows the existence of similarities as well as substantial differences in science-policy interactions among the five cases. Overall the case studies illustrate how the common ideal of natural science-based management is distorted by the practical realities of policy- and political decision-making under conditions of ecosystem complexity, uncertainty, sociopolitical ambiguity and stakeholder disagreements. We therefore argue that it is important to consider two main aspects when aiming to improve the 'governability' (Kooiman and Bavinck 2013) of marine socioecological systems like the Baltic Sea. The first aspect argues for a diversity of knowledge perspectives and suggests a balance between natural science-dominated processes and procedures and more precautionary, participatory governance approaches that take account of social dimensions and stakeholder's knowledge contributions (cf. Linke and Jentoft 2014).

The second aspect regards maintaining a balance between the ideals of adopting holistic governance approaches such as EAM and consideration of context-dependent requirements of specific societal sectors, environmental problems and risks (e.g. linked multiple policy objectives, risk types and different states of knowledge and sociopolitical dimensions).

From our analysis, we identify *five key issues and challenges* that, if adequately addressed, may improve assessment-management interactions, facilitate the implementation of EAM and thus finally improve environmental governance:

1. The *organisational structures* of science-policy interfaces need to allow for more effective, i.e. timely and context- and place-dependent, interaction between assessment activities and management responses, while simultaneously opening possibilities of distributing power, proof and responsibility to relevant actor groups, all of which would be in line with EAM. This would imply a *regional and ecosystem basis* of assessment-management interactions capable of addressing prioritised knowledge gaps and developing regional knowledge management and monitoring to strengthen region-based scientific advice.
2. The *integration of different forms and states of* (*incomplete*) *knowledge* is currently undeveloped in assessment and advice to policy- and decision-making.

This reduces possibilities of addressing the social dimension of sustainable development as well as interdependencies among environmental issues. Knowledge inclusion is therefore a topic that could potentially help to harmonise and democratise policy- and decision-making, as well as contribute to identifying and reducing existing blind knowledge spots.

3. *Stakeholder participation and deliberation* also require more careful consideration, especially in some of the cases, since the incorporation of practitioner's perspectives and local knowledge in assessments as well as stakeholder deliberation in management are vital for the successful implementation of EAM. How to arrange this in an effective, meaningful and purposeful manner in each case, however, needs to be further studied.
4. We also feel that there is substantial room for improvement with regard to *coping with scientific uncertainty and stakeholder disagreements* in both assessment and management. We conclude that there are examples of science-based precautionary approaches and methods but that a comprehensive and coherent strategy for addressing uncertainty and disagreements is often lacking. Both the institutional designs of the investigated science-policy interfaces as well as the processes of stakeholder interaction need to be adapted to meet the intentions of 'good' environmental governance as laid down in various EU principles (e.g. in the CFP, EC 2002).
5. Despite the commonly expressed idea of a clear separation between assessment and management, the studied science-policy interactions appear often rather *diffuse and politicised*. This *lack of transparency and reflection* about the practical realities of science-policy interactions and how they evolve and are constructed may mislead both political decision-makers and the public and thus potentially hamper effective management progress. As it is unavoidable that the spheres of science and policy are blurred, it is even more important to be transparent about points of departures, underlying values, methodological choices and approaches used in assessment and management related to environmental governance. Furthermore, improved communication about the internal processes of science and policymaking and their interlinkages ('co-production') is similarly important to counteract existing governance deficits.

Acknowledgements This chapter draws on results from the research programme 'Environmental Risk Governance of the Baltic Sea' (2009–2015, Michael Gilek, programme coordinator), which involved research teams from Södertörn University in Sweden, Åbo Akademi University in Finland, Dialogik/Stuttgart University in Germany and Gdansk University in Poland. The funding came from the Foundation for Baltic and East European Studies and the European Community's Seventh Framework Programme (2007–2013) under grant agreement no. 217246 made with the joint Baltic Sea research and development programme BONUS, as well as from the German Federal Ministry of Education and Research (BMBF), the Swedish Environmental Protection Agency, the Swedish Research Council FORMAS, the Polish Ministry of Science and Higher Education and the Academy of Finland. SL also acknowledges funding from the Swedish Research Council and Riksbankens Jubileumsfond. We wish to express our warmest thanks to these institutions for enabling us to conduct this research, to all participants in the research programme that directly or indirectly provided useful input and to all informants sharing their experiences, as well as to two reviewers for their constructive comments on an earlier version of the chapter.

References

Allanou R, Hansen BG, van der Bilt Y (1999) Public availability of data on EU high production volume chemicals. European Commission, European Chemicals Bureau, Ispra

Andersen JH, Axe P, Backer H, Carstensen J, Claussen U, Fleming-Lehtinen V, Järvinen M, Kaartokallio H, Knuuttila S, Korpinen S, Kubiliute S, Laamanen M, Lysiak-Pastuszak E, Martin G, Murray C, Møhlenberg F, Nausch G, Norkko A, Villnäs A (2011) Getting the measure of eutrophication in the Baltic Sea: towards improved assessment principles and methods. Biogeochemistry 106:137–156

Backer H, Leppänen JM, Brusendorff AC, Forsius K, Stankiewicz M, Mehtonen J, Pyhälä M, Laamanen M, Paulomäki H, Vlasov N, Haaranen T (2010) HELCOM Baltic Sea Action Plan – a regional programme of measures for the marine environment based on the ecosystem approach. Mar Pollut Bull 60:642–649

Boström M, Grönholm S, Hassler B (2016) The ecosystem approach to management in Baltic Sea governance: towards increased reflexivity? In: Gilek M et al (eds) Environmental governance of the Baltic Sea. Springer, Dordrecht

BFFE (2013) Viewpoints from the farmer organisations around the Baltic Sea to the proposal for the ministerial declaration concerning revised HELCOM Baltic Sea Action Plan (BSAP). Available from: http://www.lrf.se

Bijker WE, Bal R, Hendriks R (2009) The paradox of scientific authority – the role of scientific advice in democracies. MIT Press, Cambridge, MA

Cash DW, Clark WC, Alcock F, Dickson NM, Eckley N, Guston DH, Jäger J, Mitchell RB (2003) Knowledge systems for sustainable development. Proc Natl Acad Sci U S A 100(14):8086–8091

Casini M, Hjelm J, Molinero JC, Lövgren J, Cardinale M, Bartolino V (2009) Trophic cascades promote threshold-like shifts in pelagic marine ecosystems. PNAS 106:197–202

COM (2001) European governance: a white paper, Commission of the European Communities 2001 428 final, Brussels

Conley DJ (2012) Save the Baltic Sea. Nature 486:463–464, Comment

Conley DJ, Paerl HW, Howarth RW, Boesch DF, Seitzinger SP, Havens KE, Lancelot C, Likens GE (2009a) Controlling eutrophication: nitrogen and phosphorus. Science 323:1014–1015

Conley DJ, Bonsdorff E, Carstensen J, Destouni G, Gustafsson BG, Hansson LA, Rabalais NN, Voss M, Zillén L (2009b) Tackling hypoxia in the Baltic Sea: is engineering a solution? Viewpoint. Environ Sci Technol 43:3407–3411

Degnbol D, Wilson DC (2008) Spatial planning on the North Sea: a case of cross-scale linkages. Mar Policy 32:189–200

EC (2000) Directive 2000/60/EC of the European Parliament and of the Council establishing a framework for Community action in the field of water policy. OJ L 327:1–72

EC (2002) Council Regulation (EC) No 2371/2002 on the conservation and sustainable exploitation of fisheries resources under the Common Fisheries Policy. OJ L 358/59

EC (2003) Directive 2002/95/EC of the European Parliament and of the Council on the restriction of the use of certain hazardous substances in electrical and electronic equipment. OJ L 37:19–23

EC (2006) Regulation (EC) 1907/2006 of the European Parliament and of the Council concerning the Registration, Evaluation, Authorisation and Restriction of Chemicals (REACH). OJ L 396:1–849

EC (2008) Directive 2008/56/EC of the European Parliament and of the Council establishing a framework for community action in the field of marine environmental policy (Marine Strategy Framework Directive). OJ L 164:19–40

ECB (2003) Technical guidance document on risk assessment. Part I–IV. Joint Research Centre, European Commission, Ispra

Elofsson K (2010) Baltic-wide and Swedish nutrient reduction targets. An evaluation of cost-effective strategies. Report to the Expert Group of Environmental Studies 2010:2. Ministry of Finance, Stockholm

Eriksson J, Karlsson M, Reuter M (2010a) Technocracy, politicization, and non-involvement: politics of expertise in the European regulation of chemicals. Rev Policy Res 27:167–185

Eriksson J, Karlsson M, Reuter M (2010b) Scientific committees and EU policy: the case of SCHER. In: Eriksson J, Gilek M, Rudén C (eds) Regulating chemical risks: European and global challenges. Springer, Dordrecht

EU (2014) Regulation (EU) 1143/2014 of the European Parliament and of the Council on the prevention and management of the introduction and spread of invasive alien species. OJ L 317:35–55

Fisher E (2008) The 'perfect storm' of REACH: charting regulatory controversy in the age of information, sustainable development, and globalization. J Risk Res 11:541–563

Funtowicz SO, Ravetz JR (1993) Science for the post-normal age. Futures 25:739–755

Funtowicz SO, Strand R (2007) Models of science and policy. In: Traavik T, Lim LC (eds) Biosafety first. Tapir, Trondheim, pp 263–278

Garcia SM, Zerbi A, Aliaume C, Do Chi T, Lasserre G (2003) The ecosystem approach to fisheries. Issues, terminology, principles, institutional foundations, implementation and outlook. FAO fisheries technical paper. No. 443. FAO, Rome

Gezelius S (2008) The arrival of modern fisheries management in the North Atlantic: a historical overview. In: Gezelius S, Raakjær J (eds) Making fisheries management work. Springer, Dordrecht, pp 27–40

Gilbert N (2011) Data gaps threaten chemicals safety law. Nature 475:150–151

Gilek M, Hassler B, Jönsson AM, Karlsson M (eds) (2011) Coping with complexity in Baltic Sea risk governance. Spec Issue Int Sci J AMBIO 40(2) doi:10.1007/s1^3280-010-0122-4

Gilek M, Hassler B, Engkvist F, Kern K (2013) The HELCOM Baltic Sea Action Plan: challenges of implementing an innovative ecosystem approach. In: Henningsen B, Etzold T, Pohl AL (eds) Political state of the region report 2013 – trends and directions in the Baltic Sea region. Baltic Development Forum, Copenhagen, pp 58–61

Gilek M, Karlsson M, Udovyk O, Linke S (2015) Science and policy in the governance of Europe's marine environment – the impact of Europeanization, regionalization and the ecosystem approach to management. In: Gilek M, Kern K (eds) Governing Europe's marine environment. Europeanization of regional seas or regionalization of EU policies? Ashgate Publishing, Farnham

Gilek M, Karlsson M, Linke S, Smolarz K (2016) Environmental governance of the Baltic Sea – identifying key challenges, research topics and analytical approaches. In: Gilek M et al (eds) Environmental governance of the Baltic Sea. Springer, Dordrecht

Gren IM (2008) Costs and benefits from nutrient reductions to the Baltic Sea. The Swedish Environmental Protection Agency. Report 5877

Griffin L (2009) Scales of knowledge: North Sea fisheries governance, the local fisherman and the European scientist. Environ Polit 18:557–575

Gustafsson BG, Schenk F, Blenckner T, Eilola K, Meier M, Müller-Karulis B, Neumann T, Ruoho-Airola T, Savchuk O, Zorita E (2012) Reconstructing the development of Baltic Sea eutrophication 1850–2006. AMBIO 41:534–548

Hammer M (2015) The ecosystem management approach. Implications for marine governance. In: Gilek M, Kern K (eds) Governing Europe's marine environment. Europeanization of regional seas or regionalization of EU policies? Ashgate Publishing, Farnham

Hatchard J, Gray T (2014) From RACs to advisory councils: lessons from North Sea discourse for the 2014 reform of the European Common Fisheries Policy. Mar Policy 47:87–93
Hassler B (2011) Accidental versus operational oil spills from shipping in the Baltic Sea – risk governance and management strategies. AMBIO 40:170–178
Hassler B (2016) Oil spills from shipping: a case study of the governance of accidental hazards and intentional pollution in the Baltic Sea. In: Gilek M et al (eds) Environmental governance of the Baltic Sea. Springer, Dordrecht
Hassler B, Söderström S, Lepoša N (2010) Marine oil transportations in the Baltic Sea area. Deliverable 6 within the RISKGOV project. Available from: http://www.sh.se/riskgov
Hegland T, Raakjær J (2008) Recovery plans and the balancing of fishing capacity and fishing possibilities: path dependence in the Common Fisheries Policy. In: Gezelius S, Raakjær J (eds) Making fisheries management work. Springer, Dordrecht, pp 131–159
HELCOM (2007) HELCOM Baltic Sea Action Plan. Helsinki Commission, Helsinki
HELCOM (2010) Ecosystem health of the Baltic Sea 2003–2007: HELCOM initial holistic assessment. Baltic Sea Environmental Proceedings No. 122
HELCOM (2013) Approaches and methods for eutrophication target setting in the Baltic Sea region. Baltic Sea Environment Proceedings No. 133, Helsinki Commission, Helsinki
HELCOM (2014) HELCOM guide to alien species and ballast water management in the Baltic Sea. HELCOM, Baltic Marine Environment Protection Commission, Helskinki
Jasanoff S (ed) (2004) States of knowledge: the co-production of science and social order. Routledge, London
Jasanoff S, Wynne B (1998) Science and decision-making. In: Rayner S, Malone EL (eds) Human choice and climate change, vol 1, The societal framework. Battelle, Columbus, pp 1–87
Johansson S, Wulff F, Bonsdorff E (2007) The MARE research program 1999–2006: reflections on program management. AMBIO 36:1–4
Karlsson M (2006) The precautionary principle, Swedish chemicals policy and sustainable development. J Risk Res 9:337–360
Karlsson M (2010) The precautionary principle in EU and U.S. chemicals policy: a comparison of industrial chemicals legislation. In: Eriksson J, Gilek M, Rudén C (eds) Regulating chemical risks: European and global challenges. Springer, Dordrecht
Karlsson M, Gilek M, Udovyk O (2011) Governance of complex socio-environmental risks: the case of hazardous chemicals in the Baltic Sea. AMBIO 40:144–157
Karlsson M, Gilek M (2016) Governance of chemicals in the Baltic Sea region: a study of three generations of hazardous substances. In: Gilek M et al (eds) Environmental governance of the Baltic Sea. Springer, Dordrecht
Karlsson M, Gilek M, Lundberg C (2016) Eutrophication and the ecosystem approach to management: a case study of Baltic Sea environmental governance. In: Gilek M et al (eds) Environmental governance of the Baltic Sea. Springer, Dordrecht
Knudsen OF, Hassler B (2011) IMO legislation and its implementation: accident risk, vessel deficiencies and national administrative practices. Mar Policy 35(2):201–207
Kooiman J, Bavinck M (2013) Theorizing governability – the interactive governance perspective. In: Bavinck M, Chuenpagdee R, Jentoft S, Kooiman J (eds) Governability of fisheries and aquaculture: theory and applications. Springer, Dordrecht
Kortenkamp A, Backhaus T, Faust M (2009) State of the art report on mixture toxicity. Final report of a project on mixture toxicology and ecotoxicology commissioned by the European Commission, DG Environment
Lemke P, Smolarz K, Zgrundo A, Wolowicz M (2010) Biodiversity with regard to alien species in the Baltic Sea region. RISKGOV report to BONUS EEIG Programme; University of Gdansk, Institute of Oceanography, Gdynia
Leppäkoski E, Laine AO (2009) Alien species. In: Zweifel UL, Laamanen M (eds) Biodiversity in the Baltic Sea. An integrated thematic assessment on biodiversity and nature conservation in the Baltic Sea. Baltic Sea Environmental Proceedings No. 116B. HELCOM – Baltic Marine Environment Protection Commission, Helskinki, pp 119–124

Lidskog R (2014) Representing and regulating nature: boundary organisations, portable representations, and the science-policy interface. Environ Polit 23:670–687
Linke S, Bruckmeier K (2015) Co-management in fisheries – experiences and changing approaches in Europe. Ocean Coast Manag 104:170–181
Linke S, Jentoft S (2013) A communicative turnaround: shifting the burden of proof in European fisheries governance. Mar Policy 38:337–345
Linke S, Jentoft S (2014) Exploring the phronetic dimension of stakeholders knowledge in EU fisheries governance. Mar Policy 47:153–161
Linke S, Dreyer M, Sellke P (2011) The Regional Advisory Councils: what is their potential to incorporate stakeholder knowledge into fisheries governance? AMBIO 40:133–143
Linke S, Gilek M, Karlsson M, Udovyk O (2014) Unravelling science-policy interactions in environmental risk governance of the Baltic Sea: comparing fisheries and eutrophication. J Risk Res 17(4):505–523
Long R (2010) The role of Regional Advisory Councils in the European Common Fisheries Policy: legal constraints and future options. Int J Mar Coast Law 25:289–46
LRF (2010) Synpunkter från LRF på Slutrapport av Naturvårdsverkets regeringsuppdrag 17 i regleringsbrev för 2008: Förslag till nationella åtgärder enligt Baltic Sea Action Plan (BSAP) (in Swedish). [Opinions from LRF on the final report of the Swedish Environmental Protection Agency's government report 17/2008: Recommendations for national measures according to Baltic Sea Action Plan]
LRF (2013) Hur återställer vi Östersjön? Effektivare strategier mot ett mindre övergött hav (in Swedish). Lantbrukarnas Riksförbund, Stockholm
Nielsen KN, Holm P (2008) The TAC machine: on the institutionalization of sustainable fisheries resource management. In: Nielsen KN (ed) Science|Politics: boundary construction in mandated science – the case of ICES' advice on fisheries management. PhD dissertation, University of Tromsø
Nielsen KN, Holm P, Aschan M (2015) Results based management in fisheries: delegating responsibility to resource users. Mar Policy 51:442–451
Pihlajamäki M, Tynkkynen N (eds) (2011) Governing the blue green Baltic Sea – societal challenges of marine eutrophication prevention. FIIA REPORT 31
Renn O (2008) Risk governance: coping with uncertainty in a complex world. Earthscan, London
Rousi H, Kankaanpää H (eds) (2012) The ecological effects of oil spills in the Baltic Sea – the national action plan of Finland. Environmental Administration Guidelines 6en 2012. Finnish Environment Institute, Helsinki
Selin H (2007) Coalition politics and chemicals management in a regulatory ambitious Europe. Glob Environ Polit 7(3):63–93
Sellke P, Dreyer M, Linke S (2016) Fisheries: a case study of Baltic Sea environmental governance. In: Gilek M et al (eds) Environmental governance of the Baltic Sea. Springer, Dordrecht
Smolarz K, Biskup P, Zgrundo A (2016) Biological invasions: a case study of Baltic Sea environmental governance. In: Gilek M et al (eds) Environmental governance of the Baltic Sea. Springer, Dordrecht
Stigebrandt A, Kalén O (2013) Improving oxygen conditions in the deeper parts of Bornholm Sea by pumped injection of winter water. AMBIO 42:587–595
Stigebrandt A, Rahm L, Viktorsson L, Ödalen M, Hall P, Liljebladh B (2014) A new phosphorus paradigm for the Baltic proper. AMBIO 43:634–643
Stirling A (2007) Risk, precaution and science: towards a more constructive policy debate. EMBO Rep 8:309–315
Stirling A (2008) "Opening up" and "closing down" – power, participation, and pluralism in the social appraisal of technology. Sci Technol Hum Values 33:262–294
Stirling A (2010) Keep it complex. Nature 468:1029–1031
Urquhart J, Acott T, Symes D, Zhao M (eds) (2014) Social issues in sustainable fisheries management. Springer, Dordrecht
Weingart P (1999) Scientific expertise and political accountability: paradoxes of science in politics. Sci Public Policy 26:151–161

Wilson DC (2009) The paradoxes of transparency: science and the ecosystem approach to fisheries management in Europe. Amsterdam University Press, Amsterdam

Wulff F, Savchuk OP, Sokolov A, Humborg C, Mörth CM (2007) Management options and effects on a marine ecosystem: assessing the future of the Baltic. AMBIO 36:243–249

Zavaleta ES, Hobbs RJ, Mooney HA (2001) Viewing invasive species removal in a whole-ecosystem context. Trends Ecol Evol 16:454–459

Chapter 9
Risk Communication and the Role of the Public: Towards Inclusive Environmental Governance of the Baltic Sea?

Anna Maria Jönsson, Magnus Boström, Marion Dreyer, and Sara Söderström

Abstract This chapter focuses on forms of and challenges for risk communication within regional environmental governance, based on an analysis of five environmental risks in the Baltic Sea – marine oil transportation, chemicals, overfishing, eutrophication and alien species. We address questions about how risks are framed and communicated and also analyse the role of communication in the governance process. Our main focus is on risk communication with the public (e.g. existing institutional arrangements and procedures of risk communication), but we also relate this analysis to discussions on communication with a broad range of actors and issues of stakeholder participation and communication. In the study we have identified some examples of relatively well-working risk communication with parts of the organised public in the Baltic Sea region (BSR), such as in fisheries or eutrophication, but also a number of different barriers and obstacles. Our key result from this study is that BSR consists of many national institutions for risk communication, but that there are hardly any centralised institutions for risk communication activities relating to environmental governance in the region. Another key conclusion is that public risk communication in this array of cross-national environmental risks is restricted mainly to (one-way) information. Against this backdrop and from our

A.M. Jönsson (✉)
School of Culture and Education, Södertörn University, 14189 Huddinge, Sweden
e-mail: anna.maria.jonsson@sh.se

M. Boström
School of Humanities, Education and Social Sciences, Örebro University, SE-701 82 Örebro, Sweden
e-mail: magnus.bostrom@oru.se

M. Dreyer
DIALOGIK, Lerchenstrasse 22, 70176 Stuttgart, Germany
e-mail: dreyer@dialogik-expert.de

S. Söderström
School of Natural Sciences, Technology and Environmental Studies, Södertörn University, 14189 Huddinge, Sweden
e-mail: sara.soderstrom@sh.se

M. Gilek et al. (eds.), *Environmental Governance of the Baltic Sea*, MARE Publication Series 10, DOI 10.1007/978-3-319-27006-7_9

empirical and theoretical knowledge of risk communication and the role of the public, we finally suggest some ways for improvement.

Keywords Stakeholder participation • Ecosystem approach to management • Public involvement • News media • Framing

9.1 Introduction and Background

9.1.1 Background

Trans-boundary environmental risks that affect us all over the globe have raised a need for new political spaces, political identities and the emergence of a global civil society (Castells 2008). This transformation of society has among other things made the concept of governance relevant for describing a new situation for dealing with environmental policies and risks, involving different actors at different levels. Despite several initiatives to counteract negative trends, it is clear that human activities still cause severe environmental problems and risks which means that structures and processes for environmental governance have to be improved (cf. Gilek et al. 2011). According to Renn (2008), risk governance consists of four phases: pre-assessment, appraisal, evaluation and management. Risk communication is in different ways included in all of these and therefore can be considered as a key component of risk governance. In this chapter, we focus on institutionalised forms of risk communication and especially on risk communication with and to the public with regard to environmental governance of the Baltic Sea.

The implementation of the ecosystem approach to management (EAM) has been identified as a key means for achieving a healthy marine environment. For example, in the Baltic Sea region (BSR), environmental risk governance strategies have so far not been designed to deliver the holistic appraisals and integrated management that would be required by this approach (Dreyer et al. 2011; Hassler et al. 2013; HELCOM 2010; McGlade 2010). The application of EAM in BSR has been a central pillar of HELCOM's Baltic Sea Action Plan (BSAP) from 2007 (HELCOM 2007). EAM is identified as a means to address major pressures affecting the Baltic Sea marine environment in a more integrated and effective manner (cf. Karlsson et al. 2011). The EU has also endorsed EAM in different directives (Dreyer et al. 2011), and previous research has shown that deliberation processes are also vital for the implementation of it (Dreyer et al. 2014).

In this chapter, we discuss forms, opportunities and challenges for risk communication and the role of the public in implementing EAM in environmental and regional governance, using BSR as a case. Besides being an ecologically and politically highly diverse and important region to study in itself, BSR is held forward as a pioneer for introducing new forms of environmental governance (cf. Joas et al.

2008). Environmental governance of the Baltic Sea is not only a case of regional governance but also linked to the area of marine governance (see, e.g. van Leeuwen and Kern 2013). Even if BSR is a particular case, the RISKGOV project (cf. Introduction in this book) has contributed knowledge relevant for the wider fields of regional, marine and environmental governance. We argue, accordingly, that results presented in this chapter are relevant for the understanding of (public) risk communication, within a cross-country governance context, in general.

Various actors (e.g. policy-makers, social scientists, etc.) agree that for societies to be able to manage and govern global risks, there is a need for improved transnational communication, more harmonised decision-making and multi-stakeholder participation as well as for the increased involvement of citizens through information and dialogue. In the policy-making context, this is described as 'good governance' (cf. Whiteside 2006). For example, the Aarhus Convention (1998) emphasises the role of public deliberation in environmental decision-making. So too do the EU directive on Environmental Impact Assessment (EIA), the EU Sixth Environmental Action Programme (European Commission 2001a) [1] and the European Commission's White Paper on European Governance (European Commission 2001b). From a theoretical perspective, 'good governance' can also be defined as 'reflexive governance' and includes aspects like transparency, broad participation and self-scrutiny (see, e.g. Boström et al. 2016; Hassler et al. 2013; Voss and Kemp 2006).[2] According to Oran Young, 'governance' is a wider term than government, as it is 'a social function centred on efforts to steer or guide societies toward collectively beneficial outcomes and away from outcomes that are collectively harmful' (2009: 12). This function often includes governments playing a role but is much wider comprising other actors from other social spheres as well. Governance is, accordingly, not restricted to hierarchical and top-down relations but also involves citizens and stakeholders in network-like constellations. The communication model underpinning this mode of governance is 'dialogue' (e.g. Felt and Fochler 2010; Jentoft et al. 2007; Pierre and Peters 2000).

In a previous article about stakeholder participation and communication in relation to environmental issues in BSR, Dreyer and colleagues (2014) concluded that it is crucial to acknowledge the role of the public, enhance efforts to communicate EAM and raise public awareness of and interest in EAM, for a successful governance process and implementation of EAM. Increased public debate would also help to prepare the ground for 'reflexive governance' based on a deliberative ideal. At an aggregated level, the public and citizens are important counterparts to policymakers as 'public opinion', voters and grassroots movements.

[1] There is now also a seventh action programme that however does not mention public deliberation among the main priorities, only 'better information' http://ec.europa.eu/environment/newprg/index.htm

[2] See Boström et al. (2016) for an analysis of BSR environmental governance using the theory of reflexive governance.

Dreyer and colleagues furthermore concluded that it is a huge challenge to organise such transnational citizen involvement in an effective and fair manner. As mentioned above, together with participation, increased transparency in decision-making processes is vital for reflexive governance. In practice however this transparency is seen in relation to well-organised stakeholders and representatives of different organisations and generally does not include the common citizen.

Our main argument in this chapter is that risk communication to the public (particularly in its dialogic form) is a necessary element in reflexive governance, which in turn is a key requirement for serious implementation of EAM and a move towards a more sustainable marine environment. We also argue that wide public acceptance of general risk policies and their underlying principles is a necessary condition in order to successfully implement such measures and that there is a need to focus on the role of the public. To better understand the forms and consequences of public involvement and risk communication, this chapter analytically focuses on institutional arrangements and arenas for public risk communication with a main focus on the news media given its ability to influence frames and agendas. These are some of the main factors that shape the relationship between the general public and policymakers in the field of political communication (cf. Mazzoleni and Schulz 1999; Schulz 2004; Strömbäck 2008).

9.1.2 Aim and Questions

Against this background, the overarching *purpose* of this chapter is to describe and discuss conditions for risk communication in relation to the five main environmental risks in BSR, with a particular focus on communication with and to the larger public. We also suggest possible pathways to encourage aspects of risk communication that facilitate reflexive governance and implementation of EAM. We will empirically explore the following questions:

1. What are the existing institutional arrangements and procedures of risk communication at the regional (Baltic Sea) level? What forms of communication can we identify (linear vs. dialogue, formal vs. informal, etc.)?
2. What is the role of public risk communication in environmental risk governance of BSR, and how can this communication be understood and characterised?
3. What role can be ascribed to different kinds of news media? What is the role of framing, arenas and agendas?

The empirical data used for analysis in this chapter is drawn from five case studies, which were conducted in the RISKGOV project. The case studies analyse the governance of the marine environmental risks of eutrophication (Haahti et al. 2010), invasive alien species (IAS) (Lemke et al. 2010), overfishing (Sellke et al. 2010), oil discharges (Hassler et al. 2010) and chemical pollution (Udovyk et al. 2010) with respect to three governance dimensions, one of which is 'processes of stakeholder communication'. A common analytical framework guided all five case studies ensuring the comparability of the cases, which were performed using similar research designs (cf. Gilek et al. 2011). All case studies employed an explorative interpretative approach of qualitative social research, and their results have been derived from three main sources: document analysis, semi-structured qualitative interviews and a series of round-table events. This allows us to draw conclusions about communicative aspects of environmental governance of BSR, whereas we will also when possible use the case of BSR to discuss regional and marine environmental governance more generally.

In the next section, we will introduce our main theoretical perspectives that focus on risk communication in environmental governance, risk communication with and to the public and the role of the news media. We will then turn to our analysis of the empirical material and start with a discussion about existing institutional arrangements and procedures of risk communication and the forms of communication that can be identified. Thereafter we address the topic of risk communication with the public and how this communication can be understood and characterised with a special focus on arenas for communication and the role of the media. Finally, we conclude the chapter with a discussion on the role of public risk communication in (regional) environmental risk governance and possible ways forward.

9.2 Theoretical Perspectives

9.2.1 Environmental Governance and Risk Communication

We use the concept of risk communication as defined in the 1989 report on 'Improving Risk Communication', produced by the Committee on Risk Perception and Communications of the US National Research Council (NRC). The report argues that risk communication is:

> [...] an interactive process of exchange of information and opinion among individuals, groups and institutions. It involves multiple messages about the nature of risk and other messages, not strictly about risk, that express concerns, opinions or reactions to risk messages or to legal and institutional arrangements for risk management. (US NRC 1989: 1)

Risk communication is a fundamental part of environmental risk governance (cf. Renn 2008), but can also be seen as part of the wider notion of environmental communication. Environmental communication by Cox is defined as 'the pragmatic and constitutive vehicle for our understanding of the environment as well as our relationships to the natural world; it is the symbolic medium that we use in constructing environmental problems and negotiating society's different response to them' (Cox 2010: 20). Contemporary risks are rarely a first-hand experience, but something communicated through different channels of information flows in today's modern society. Risk communication is a field of research that addresses communication between experts and the public, or as it has traditionally been, risk information given to the public by experts. Renn (2008) identifies four different types of risk communication: documentation, information, dialogue and involvement.

Communication as it is used in this context thus includes on the one hand the provision of information (which in some respects also includes documentation) and on the other hand dialogue and involvement. These two forms are like two different models of communication (see, e.g. Fiske 1990). It is relevant to distinguish between them in analyses of governance activities even if there are obvious overlaps between the two.

In general, research in risk communication still rests on a traditional, and in some sense obsolete, notion of communication as a linear process, which might be successful but could also fail. This notion has its roots in the classical source-receiver model where a message is sent from a source to a transmitter who decodes it and sends it towards the receiver, the audience (Cox 2010; Fiske 1990). The model has been applied and somewhat amended to fit patterns of science and risk communication; where the *sources* consist of, for example, scientists, public agencies, interest groups and/or eye witnesses; where the *transmitters* are mass media, public institutions, interest groups and opinion leaders; and where the *receivers* include the general public, specific target audiences, group members or exposed individuals. Traditionally risk communication studies have had a focus on the means of transmission as well as on the composition of the message and the ability to persuade the audience in a particular way with the main aim of changing audience behaviour (Breakwell 2007). Traditionally the communication of risk messages thus has mainly been about experts 'informing' the public (e.g. with regard to the public's 'right-to-know').

The view of seeing communication solely as a one-way endeavour (from 'experts' to 'laymen') has in many ways been abandoned on behalf of a more inclusive and process-oriented view of communication. Here, communication is seen as evolving through systems and networks over time and within deliberate decision-making about risks (Breakwell 2007; Renn 2008). This linear model has been criticised for '…its naive conception of the public. It views them as passive recipients of information, taking no account of how the information they receive will interact with their pre-existing knowledge and attitudes and ignoring any demands they may have for what they learn to be relevant to their individual situation' (Gregory and Miller 2000: 97).

Public participation and its conceptualisation in risk policy have thus shifted from a technical one-way communication towards a more inclusive approach, in which there is a 'two-way learning' between science and society (Pidgeon et al. 2006). Here we see a parallel development in the field of science communication with a shift from the deficit model with its focus on *informing* the public for the purpose of increasing scientific knowledge among citizens and thus fostering more positive attitudes towards science to a more dialogical ideal, expressed in Bucchis' ideas on avoiding the transfer paradigm and instead seeing communication as a 'crosstalk' and a 'double helix' with one strand representing scientific discourse and the other representing public discourse (Bucchi 2004).

9.2.2 Why Involve the Public?

From the perspective of the so-called civic science, citizens and the public have a stake in the science-politics interface, an interface that can no longer be viewed as an arena for scientific experts and policy-makers only (Bäckstrand 2003). So what do we mean by the 'public'? In governance theory the public is generally framed as *concerned* publics, meaning those affected by decisions, or what Beck (1992:61) describes as 'the voices of the side effects'. Models of deliberative democracy and the public sphere however refer to citizens in general and the notion of the public interest (Dahlgren 2005; Habermas 1989). Both these perspectives point to the difficulty of defining what is at stake for the public – i.e. what is 'the public interest' – and who represents the 'general citizen'. Researchers working on the public/science/policy interface point, for example, to problems with participatory processes in relation to the question of whom the public actually represents and the type of people interested in participating (Janse and Konijnendijk 2007). In this chapter, we consider stakeholders as organised interest groups and the public as unorganised, representing the individual citizen.

Risk communication with the public implies taking into account the values, knowledge and experience of citizens. Such inclusion can improve governance and also make decisions more legitimate, robust and easier to implement (see, e.g. Stirling 2005; Whiteside 2006). Public involvement affects how problems and solutions are identified and defined since participants can have a great influence on how issues are framed. In the fields of science and risk communication, much attention has been directed towards the role of deliberative activities and processes for public participation and communication. Such activities include public hearings, conferences and advisory groups (see, e.g. Brake and Weitkamp 2009; Hagendijk and Irwin 2006; Lidskog et al. 2010).

A major incentive for activating public dialogue processes is to restore the public's confidence in government authorities including risk regulators, although many argue that increased public communication and a democratised scientific process are also good things in themselves. Increased public involvement may reduce, for example, mutual distrust. However, certain drawbacks of public involvement may appear. Increased transparency and public debate regarding complex and uncertain risk issues may result in increasing worries and larger unpredictability of risk perception and behaviour (cf. Frewer 2003). This in turn could be problematic for an efficient governance process (cf. Irwin 2008).

According to Ortwin Renn, to communicate risk to the public is a difficult task (Renn 2008). Risk communication comprises several essential dimensions including education/enlightenment of the audience and risk training for involved parties to cope with the risks and to build up confidence in risk management as well as to help create confidence towards risk handling institutions, it also enhances cooperation and conflict resolution in risk-related decision-making processes. These specific functions of risk communication all require slightly different forms of communication, including documentation, information, mutual dialogue and involvement. In risk communication the issue of credibility is of key importance and so too the ability to catch those who are potentially interested (Renn 2008).

9.2.3 Risk Communication and the Public Sphere

To acknowledge the role of the public in risk governance and communication is at least a twofold endeavour. On the one hand, it is about including the public as citizens (organised and individually) in governance processes via, for example, hearings. On the other hand, it is about raising awareness and making these issues a prominent part of the public discourse putting them on the public agenda. Political decisions are widely affected by the relationship between the agendas and frames of policy-makers, the public and the media (cf. Asp 1986). The news media and public debate are essential for communication between policy-makers and the public and for creating common agendas and framing risks. Over the last decade, so-called social media like Twitter and Facebook have also been a tool for policy-makers to communicate with and inform the general public.

The media is also an arena for public representation and for different forms of participation (such as submitting letters to editors or writing opinion pieces), not least when it comes to environmental risk issues (cf. Egan Sjölander and Jönsson 2012). Media and journalism studies have highlighted that the news media generally articulates an elitist discourse (see, e.g. Shoemaker and Reese 1996), something that is the case with regard to environmental news as well. Spokespersons from government and industry dictate the discourse of environmental news, while the views of 'ordinary' citizens or 'side effects' (cf. Beck 1992) are much more rare (cf. Cox 2010). Research also shows that those who appear in the news are able to influence on how a problem is framed in terms of causes and solutions (see, e.g. Entman 1993).

The concept of framing has roots in both psychology and sociology and in the work of sociologist Erving Goffman. Goffman discusses framing as an interpretive framework that helps individuals to process information (Goffman 1974; Pan and Kosicki 1993). In the field of policy-making theory and political sociology, the concept of framing is often used to analyse how actors are actively involved in debating, defining and setting a particular agenda and furthering its implementation (Rein and Schön 1993).

The concept of framing can be used in different kinds of analyses and has been applied in studies of stakeholder participation and communication within governance processes (see, e.g. De Marchi 2003; Dreyer et al. 2014; Jönsson 2011). Risks are inherently difficult to communicate as different interpretations and implications are bound to emerge. Framing is the work of defining and answering questions like: What is at stake? What is the risk? What is the cause and effect (Entman 1993; Schön and Rein 1994)?

Framing is an essential component in all phases of risk management (while perhaps particularly important in the pre-assessment stage; see Linke et al. 2011).

Framing denotes processes in which actors deal with social and ecological complexity. Stakeholders develop their arguments through frames, and these frames also help them find common ground for negotiations and compromises. Policy-making processes may stimulate rich debates and reflection both within and across frames, making stakeholders and the general public able to develop arguments, debate and reflect critically on policy statements. Used reflectively, frames such as the ecosystem approach, biodiversity, overfishing, precaution, sustainability and many others are useful for perceiving and understanding sets of problems in novel ways. The multilayered character of these frames may open up interpretative flexibility (cf. Klintman and Boström 2008). Such flexibility provides both barriers and bridges to communication.

Framing as such can also be viewed in light of communication and media texts. Robert Entman (1993: 52), doing exactly that, defines framing in the following manner: 'To frame is to select some aspects of a perceived reality and make them more salient in a communicating text, in such a way as to promote a particular problem definition, casual interpretation, moral evaluation, and/or treatment recommendation for the item described'. In this setting, frames can be said to define the problem, diagnose the causes, make moral statements and also suggest solutions to the problem at hand, even though a frame does not necessarily have to include all elements (Entman 1993). Media framing functions as a way to construct a specific environmental issue and elicit a response or conclusion from the media recipients (Hansen 2010).

In the next section, we will present our empirical findings and analyse risk communication in relation to our selected five environmental risk issues relevant to the Baltic Sea.

9.3 Results: Risk Communication in BSR

9.3.1 Institutionalised Risk Communication

The main finding from our cases is that there are few, if any, actual examples of systematic institutionalised public risk communication in BSR. There are also a lack of regional and transnational networks and communicative structures for information sharing with and involvement of the public. However, there are some exceptions as discussed below.

In many cases, HELCOM is the main actor in risk communication. Among other things, this organisation serves as a communicative platform and meeting place for different actors and interests (Hassler et al. 2013; Van Deever 2011). In the eutrophication case, HELCOM, with its Baltic Sea Action Plan, is seen as one of the main players (Haahti et al. 2010). The structure within HELCOM makes it possible for stakeholders such as NGOs to communicate and contribute knowledge. According to the case study report on chemical pollution, the EU and HELCOM are platforms of communication for politicians and authorities, but not for independent single scientists or laboratories (Udovyk et al. 2010). The European Environmental Agency (EEA), the European Green Spider Network (consisting of communication officers from national environmental ministries and agencies in Europe and supported by the DG Environment of the European Commission) and IMPEL are other organisations or networks that aim to distribute and communicate information regarding environmental risks in, for example, the BSR.

There is of course also communication within and among the different organs of the EU. However, the complexity of the EU creates difficulties for communication inside the organisation. This is, for example, noted in the eutrophication case study where communication between authorities at the national level is less about institutionalised formal communication, but instead about informal communication and personal contacts and networks, created through a tight interplay between science and policy (Haahti et al. 2010; Linke et al. 2014). This reliance on informal communication and personal contacts is also noted in the case of hazardous chemicals, as illustrated in the following quote from an official from a Swedish state authority: 'It is important to know the right people and have a network of contacts. If you send a letter to the agency, you will never know where it will end up' (Udovyk et al. 2010: 48). These personal contacts are not often easily applied across the entire BSR (Udovyk et al. 2010).

The lack of communication and tools and platforms for successful communication is often identified as a barrier to risk management and implementation. An example of this is that of IAS. According to the case study report on IAS, cooperation and communication between the key stakeholders was highly unsatisfactory. The weak communication between the main players, together with the lack of public debate, was identified as the main reason for low public awareness on the IAS issue (Lemke et al. 2010).

Overfishing is the risk area with the most established forms of regional risk communication. The communication takes place through Regional Advisory Councils (RACs), like the one established for the Baltic Sea in 2006 (cf. Linke et al. 2011). RACs are advisory institutions set up by the EU Commission with representatives from the fishing industry and different environmental groups (NGOs) and are often put forward as an innovative example of regional stakeholder participation (Linke and Jentoft 2013; Sellke et al. 2010). If broad inclusion is the ideal for risk governance, it is necessary to not have a static idea of who stakeholders are. A static view of stakeholders presupposes what issues are at stake and who is to be seen as holding a stake in that issue. In fact, the very definition of who the legitimate stakeholders are is part of the framing process. Governance structures, including organisations and institutions, shape these framing processes.

Officials of the German Fisheries Association stressed during an interview that the BS RAC is a forum for 'entering into real dialogue with other stakeholders, scientists and Commission officials' (Sellke et al. 2010: 29). We found that the most advanced form of regionalisation among our five case studies was in the area of fishing with RACs as the main example of multi-stakeholder regionalised advice procedures. There is however a critique directed towards the RAC system both from stakeholders themselves (environmental NGOs complain about the power asymmetries that benefit the fishing industry) and among scholars. Linke et al. (2011) suggest that a basic problem with RACs is a mismatch between the participation purpose (knowledge inclusion) and the stage in the governance process at which RACs are formally positioned (evaluation of management proposals). Their conclusion is that if the aim is to broaden the knowledge base of fisheries management, stakeholders need to be included earlier in the governance process (cf. Linke et al. 2011).

9.3.2 Forms and Platforms for Public Risk Communication

As mentioned above, Renn (2008) identifies four different forms or kinds of risk communication: documentation, information, dialogue and involvement. We find examples of all of these in our cases. In the cases of oil transportation, overfishing and eutrophication, communication can be characterised as one way from authorities to a wider public. It appears in the form of information such as statements, press releases and reports. Two-way communication and dialogue are sometimes established via different projects, which often include organised stakeholders as partners but not links to the general public. For example, in the case of eutrophication, research projects such as the Baltic COMPASS (Comprehensive Policy Actions and Investments in Sustainable Solutions in Agriculture in the Baltic Sea Region) and Baltic DEAL (Putting Best Practices in Agriculture into Work) involve farmers in BSR in the research process (see Haahti et al. 2010).

There are different media and tools for communication to and with the public, for example, eco-labels, documents, reports, articles, laws and regulation, meetings and seminars. Our results point to the preference for using digital media platforms and webpages. Different actors however use different forms of communication. Whereas scientists use reports and articles, NGOs focus more on the use of the Internet and the organisations' webpages to inform the public. Through these channels, NGOs provide information, discussion forums and ideas on what the individual citizen can do to prevent or mitigate Baltic Sea-related problems.

Awareness raising constitutes a key part of HELCOM's communication policy. Its website states that 'An essential objective is to raise general public awareness of the Baltic Sea and HELCOM actions' (quoted in Dreyer et al. 2011). HELCOM also uses its website to communicate to a wider audience. Through its website HELCOM provides a large amount of publicly accessible documentation and information on the five cases. The information formats used, for example, indicator fact sheets and thematic reports, are primarily produced for targeted users at national and Baltic-wide levels. The website mainly meets the information needs of those who already express an interest in marine environmental risk governance. This is also the case with RACs in the fisheries case and relates to the question of who is framed and considered as a ‘legitimate’ stakeholder. However, HELCOM also uses other forms of communication that can better reach the interested citizen or consumer. These include information brochures, videos and television and radio series,[3] all with the potential to raise the awareness of those who have as yet not taken an interest in the issues so far (Dreyer et al. 2011).

A significant amount of documentation and information related to environmental issues at national and EU levels is made available to the public through the websites of responsible authorities. Some websites also provide more specific information. Take for example actors dealing with chemicals like the Swedish Chemicals Agency (KEMI) that has an established and widely used website with a large number of databases and up-to-date information. In the area of fisheries, industry representatives have designed a website (www.portal-fischerei-portal.de) that aims to deliver information to the public. This is, according to the interviewees, mainly a response to the media framing of overfishing (Sellke et al. 2010). A wealth of information – from general to highly detailed, much of this related to risk assessment matters – is made available on the website of the International Council for the Exploration of the Sea (ICES). The case studies in our project do not provide examples of any websites of responsible authorities that attempt to respond to the particular concerns of the general citizen. This is different from risk communication related to, for example, food safety issues (cf. Dreyer et al. 2011).

WWF employs a traffic light system as a popularised way of informing the public (or in this case the consumers) about the sustainability practices related to particular fishes sold in the supermarket. Also green marketing, certification and labelling are used as tools for bringing simple and concise types of information to the public. A special form of communication, which is mainly used in business-to-business relations or business-to-authority relations used in the case of hazardous chemicals, is the so-called MSDS (Material Safety Data Sheets).

Actors such as industries and NGOs (with enough resources) also engage in lobbying activities in the EU and national forums for decision-making. NGOs can themselves work as communicative brokers between other actors. There are a number of research projects concerning governance of risks in BSR. To the extent that these projects include researchers and other stakeholders, they can be seen as communicative platforms for deliberation and a meeting point for different actors.[4] However, these activities seldom include the ordinary citizen.

[3] See, for example, the HELCOM radio series ‘The Baltic – a sea of change’ and the TV series ‘The Baltic – sea of surprises’, available at the HELCOM interactive webpage (http://agripollute.nstl.gov.cn/MirrorResources/10221/index.html).

[4] The RISKGOV project, for example, organised three round tables where researchers met with policy-makers, NGO representatives and scientists (reports from these round tables can be found in the RISKGOV project homepage www.sh.se/riskgov).

9.3.3 The Role(s) of the Public

Our results clearly show that actors differ in their views on public risk communication. While some consider involvement of the public fundamental, others underline the importance of expert knowledge. These positions partly reflect different national cultures and thus most likely reflect different ideals of democracy and the role of the public in different countries. The role of the public is, for example, often underplayed by interviewees from Russia, who instead underline the importance of expert knowledge (Udovyk et al. 2010). It is mostly NGOs who highlight the importance of communication and public involvement, something that is particularly evident in the overfishing case (Sellke et al. 2010) and in the case of hazardous chemicals (Udovyk et al. 2010). It could be argued that NGOs, in general, build on the idea of public information and involvement.

From our case studies, it is also obvious that the public generally is not involved or addressed in the risk assessment phase but rather (if at all) in the risk management or implementation phase. Risk assessment generally does not include dialogue with the public, and the communication phase is mainly about informing the public. Take for example the overfishing case, which otherwise has the most advanced system for stakeholder interaction: 'Overall, interviewees did not see a need for more specific public participation within the governance process. Communication with the public was mainly seen in terms of information providing' (Sellke et al. 2010: 32). The importance of informing the public – often framed in terms of the public's 'right-to-know' (RTK) – has received increased attention in risk governance processes of BSR. In the case of information provision regarding chemicals, the pollutant release and transfer registers can be seen as examples of legally binding implementation of RTK (Udovyk et al. 2010).

Risk assessment has become an important tool for informing risk managers and the general public about the different options for protecting public health and the environment (Udovyk et al. 2010). Some of the interviewees, for example, from DG Mare, saw the different national parliaments as representatives of the public, emphasising that there is no need for further public involvement in the governance process (cf. Sellke et al. 2010).

A main finding is that the public is mainly addressed and thought of in their role as consumers and less so as in their role of political citizens. For example, several actors in the chemical case, in the overfishing case and in the IAS case see the responsibility of citizens in terms of them being consumers. In the case of hazardous chemicals, consumers are often portrayed as responsible for speeding up the process of change (Udovyk et al. 2010). It appeared from our interviews that the topic of 'sustainable consumption' is gaining importance and becoming a central part of future risk communication. Today, the topic of 'sustainable consumption' seems to be of most importance in public risk communication in the overfishing case (Sellke et al. 2010). Food/eco-labels like MSC have become highly influential in the last decade.

The public as consumers are, however, thought of in different ways in the five cases. According to Agenda 21 that addresses education in BSR, it is important that the work towards a sustainable society equips citizens with education and training and raises public awareness (Baltic 21 2002). However, according to our knowledge, there are no Baltic Sea regional organisations that develop and carry out targeted educational campaigns in any of the five risk issues. We are not aware of any cases where it is either discussed or specified how public awareness should be raised and whose responsibility this is.

9.3.4 The Role of the News Media

There are different barriers to public participation and communication in environmental governance and decision-making. First, there are structural factors including lack of opportunities and resources to participate. The other main problem seems to be a lack of interest from the public (Zavestoski et al. 2006), something that at least in part could be explained by differences in agenda between scientists, policy-makers and citizens. Previous studies of communication in marine governance procedures underline the importance of common agendas. Experiences of stakeholder communication and participation reveal that it is difficult to recruit citizens to participate in deliberative procedures such as hearings (RISKGOV 2011). One possible reason is that the concerned issues are not part of the public's agenda.

We see a relationship between the amount of media reporting and action taken. In Sweden and Germany, national news media have, for example, widely acknowledged the issue of overfishing, and the extent and content of reporting has induced public authorities to take corrective action through highly targeted information provision. There is however less incentive for regional organisations like HELCOM to take such actions because of the lack of reporting and a common agenda at the regional (i.e. transnational) level (cf. Dreyer et al. 2011; Jönsson 2011). This lack of a regional public arena and agenda supported by the mass media makes it more difficult to raise awareness of regional environmental risk issues among the public of that region. The media can play an important role in the development of common agendas on a national level (cf. McCombs 2005; McCombs and Shaw 1972), which in turn is crucial for generating interest from citizens and consumers (cf. Dreyer et al. 2011). The reasons that common agendas are important are because they place a certain (risk) issue on the agenda that may consequently enhance (public) engagement and increase the possibility and quality of (public) participation, which will result in improved (risk) management and policy implementation. In this process, the media and its agenda-setting function (cf. McCombs and Shaw 1972) is of utmost importance. The basic logic behind the 'agenda-setting' concept is that there is a relationship between the amount of attention a certain issue receives in the news media and the extent to which the public consider this issue to be of special importance. This thus also means that in order to become part of the public agenda, news about environmental risks in relation to the Baltic Sea must be considered newsworthy. In today's media society, with its many communication channels and platforms, it is a challenge to reach out with messages. While agenda-setting theory mainly focuses on issues that are reported, framing in the context of media studies is about how issues are presented.

Previous studies of how news media frame environmental risks in relation to the Baltic Sea show that eutrophication together with overfishing receives the most media attention. This situation could be explained at least partly by the fact that the media focuses on stories that the consumer can identify with best (Jönsson 2011). The issue of hazardous chemicals on the other hand receives relatively little attention in the news media and is not on top of the public agenda. According to interviewees, chemical risks are framed in the media in a very narrow health-centred way related to eating contaminated fish (Udovyk et al. 2010).

Clearly, the conditions to achieve cultural resonance – the 'success' of the 'problem' – relate to possibilities of visualising the problem and reaching out to the media. The difficulty of visualising chemical pollution and the lack of media reporting were issues that interviewees mentioned (Udovyk et al. 2010). There was also less reporting on oil spills during the studied period (Jönsson 2011). Yet, oil spills can be easily visualised by the media and framed as an 'immediate threat' or 'pending danger'. However, in the case of oil spills, what is generally disseminated to the public is mostly just one side of the coin. Big accidents are always reported, whereas the more common (routinised) intentional oil spills are not (Hassler et al. 2010). This is because the media prioritises the most acute and spectacular issues, consequently resulting in the marginalisation of small catastrophes that are seen as 'normal' and not 'newsworthy' (cf. Anderson 1997; Hannigan 2006; Hansen 2010; Shoemaker and Reese 1996). Few other risk issues give rise to such intense reactions as when birds or seals are portrayed as caught in black oil, fighting for their lives. This often comes with pictures of voluntary workers removing oil from beaches and seashores. This type of catastrophe or big event framing, together with the visualisation component, can affect the kinds of societal responses that are expected of the public (cf. Hassler et al. 2010). In contrast, intentional oil spills, similar to IAS, are not issues of high priority in framing and campaigning activities of environmental NGOs, which is partly because NGOs too are constrained by 'media logic'.

Of our selected five cases, the issue of IAS appears to receive the lowest degree of media and public attention, in particular in the post-Soviet countries (Jönsson 2011; Lemke et al. 2010). IAS, in comparison to the other issues, is still new in the public's imagination in BSR (cf. Hansen 2010). Neither are environmental NGOs in general engaged in framing efforts targeted towards pushing the issue of alien species onto the public agenda nor are they mobilising support for a more developed risk policy. They do not appear to consider it a prioritised problem (Lemke et al. 2010). All in all this will affect policy outcomes. 'Since public awareness is related to political involvement and support, low awareness associated with IAS is treated by policy-makers as a public consent to neglect the issue especially in the context of non-ratified legal acts'. (Lemke et al. 2010: 55). There is accordingly no significant public pressure on, for example, HELCOM contracting parties to speed up the ratification process or consider more detailed and binding regulations.

9.3.5 The EAM and Public Risk Communication

Another general finding is that there is only very limited EAM-related risk communication targeted at the general public and that EAM has not (yet) become part of public discourse at national and transnational (BSR) levels. Public awareness of EAM is limited because there is a lack of news coverage at the regional level (Dreyer et al. 2011; Jönsson 2011). Also, the mass media seem rather unwilling to report on EAM. The reason for this is probably its low news value; it is not a thrilling or original story and in fact somewhat old-fashioned. Currently, public debate is not a likely source of pressure on stakeholders to become more engaged with EAM and its guiding principles (Dreyer et al. 2014). EAM is thus not yet, at least in BSR, part of either the public debate or agenda or part of public values and imagination. The ecological challenge is not yet a 'hot topic' for BSR as a whole. Having said that, there are differences between different risks and problem areas in terms of public attention given to them.

Although the EAM concept can potentially facilitate reflection and communication among at least some of the key actors in risk governance, it lacks essential characteristics that are necessary for it to become central to a new societal paradigm, which could push all different actors central to Baltic Sea environmental governance to jointly deliberate over and act upon it with regard to the five risk issues (Dreyer et al. 2011). Besides from being rather invisible in public discourse, EAM is used differently and of a different level of importance to the five risk cases. In, for example, the oil spill case, it was reported that some of the key actors interviewed had never encountered the term before (Hassler et al. 2010). While many or even most interviewees seem to welcome the EAM concept, it is certainly less clear if this frame helps in the development of mutual understandings of risk. An interesting finding was found in the case of overfishing, where NGO representatives claimed that communication actually could be improved by EAM since it creates a common agenda and discourse (Sellke et al. 2010). Of course, it is also necessary to define what EAM really is while framing the problem and whether or not it is a 'utopian vision', as some interviewees claim, or a fruitful way forward that alone can help achieve the goal of sustainable development, as claimed by some. All in all, the abstract and complex nature of EAM appears to be a barrier to its adoption and communication across various actor groups. Governance of environmental risks and implementation of EAM require balancing different and often conflicting interests and values. Shared frames and agendas would, for example, enhance the possibility for successful risk governance. Platforms targeted at developing and implementing a more integrated approach and discursive space to address risks related to EAM are essential. These platforms and discursive spaces are needed to facilitate communication and understanding between diverse stakeholders in relation to particular risk issues and across the different sectors that are affected by these risks. Such platforms are missing in our risk cases (cf. Dreyer et al. 2011). It is also clear from our cases that risk communication in BSR today does not make much use of digital media platforms. In the digitalised network society we live in, there are a number of possibilities to 'meet' and communicate made available by different virtual platforms, even if in practice, these possibilities are not used to their full potential (cf. Jönsson 2012; Zavestoski 2006).

9.3.6 Bridges and Barriers for Public Risk Communication

It can be concluded that while the five environmental risks are all established as environmental policy issues at different political levels including that of BSR (mainly through HELCOM and the EU), they are generally poorly represented and to different degrees in public discourse in the region. A certain level of public awareness is required in order to move towards more effective regionalised environmental risk governance as aimed at by HELCOM in particular (Dreyer et al. 2011).

Governance of marine environments often involves particular challenges for participation and communication due to the fact that many marine environments include several different actors and states. In the case of BSR, some challenges for regional governance such as a lack of a historic regional identity have been identified as significant (cf. Dreyer et al. 2014). It is worth considering that the Baltic Sea is a rather heterogeneous area where countries display rather divergent historical and contemporary traits while also having different ideas about, for example, democracy and public participation (Vangas 2010).

As discussed above, risk communication can be a bridge for environmental governance and a step towards sustainability and EAM. At the same time, nonexistent or not-working risk communication can be a barrier. Common arenas and agendas for communication are of great importance for risk communication with the public, but it is also crucial that all parties involved share the same language and concepts (this is also important for frame sharing). The difference in use of frames in communication between experts and other actors has often been a barrier for (risk) communication. So has the tendency within risk governance processes to stick to a narrow scientific/technocratic framing. For example, in the case of hazardous chemicals, one expert said: 'The information is really technical and it is really hard for common people to understand what we are doing' (Udovyk et al. 2010: 50). The role of English creates a problem particularly for delegates from Russia and other eastern states. One Russian journalist says 'everything that is in English is kind of "silent" for Russia' (Udovyk et al. 2010: 48). In the area of risk communication, different laws and regulatory documents can be seen as tools for creating a common language if it establishes common objectives and a common terminology. In this context, language thus refers to both the use of, for example, English and the use and understanding of a common concept.

9.4 Conclusions and Ways Forward

That (risk) communication to the general public should be seen as an inherent element of environmental and sustainability governance forms part of the standard rhetoric of transnational and intergovernmental organisations as well as authorities at different political levels. The need for general public support for environmental information as part of environmental governance in the Baltic Sea region can therefore not be underestimated. Environmental information is seen as especially relevant when drastic measures of risk management (such as closing commercially important areas for fishing or imposing nutrient reductions to combat eutrophication) have to be implemented. Such measures normally result in protests and conflicts as well as public debate in and through the media. Wide public acceptance of overall risk policy and its underlying principles – such as those at the core of EAM – is a necessary condition to successfully implement such measures (cf. Dreyer et al. 2011).

We have identified some examples of relatively well-working risk communication with parts of the organised public in BSR, such as in fisheries or eutrophication, but also a number of different barriers and obstacles. Our analysis of five cases shows a far from ideal situation from the perspective of reflexive or good governance with regard to how risk communication is conducted. Risk communication activities are often not firmly anchored in the organisational structure of the institutions responsible for risk assessment and management and are not understood as integral components of the entire risk regulation/governance process. It is also clear that there is no real reflection on communication activities and whether the precise forms of communication and use of mediums for communication reflect the transnational nature of risks under consideration, their context and whether they arouse, or could arouse, societal concern. Our key result from this study is that BSR consists of many national institutions for risk communication, but that there are hardly any centralised institutions for risk communication activities relating to environmental governance in the region.

Another key conclusion is that public risk communication in this array of cross-national environmental risks is restricted mainly to (one-way) information. More dialogical approaches, for example, aimed at the particular needs of citizens and consumers do not play a significant role. The particular nature of the five risk issues is one reason for this lack of dialogue since they appear to provide only few opportunities for the public to contribute to risk management. We have, however, seen that in the cases of overfishing and eutrophication, the average citizen is more or less directly addressed in his/her role as consumer and encouraged to move towards sustainable consumption. These cases are, however, mainly initiatives of environmental NGOs and businesses (cf. Dreyer et al. 2011).

Against this backdrop and from our empirical and theoretical knowledge of risk communication and the role of the public, we suggest ways forward. We see several aspects that could be improved in risk communication with regard to the involvement of the general public in BSR:

1. First of all we argue that there is a need to more firmly *anchor* risk communication activities into the organisational structure of the institutions responsible for risk assessment and management. Risk communication should be understood as an integral component of the *entire* risk governance process; that is, risk communication needs to be an ongoing activity during all stages of risk assessment and risk management, from the framing of the issue to the monitoring of risk management practice, while the target group may vary from stage to stage.
2. There should be *enhanced efforts to communicate EAM to the general public* by using a more *proactive approach.* Currently, EAM is generally perceived of as a rather abstract and technical concept that is unlikely to become part of the wider public agenda. In the context of implementing BSAP, HELCOM (and also public authorities at national levels) increased their efforts to communicate the EAM concept to journalists and other disseminators through different modes of popularisation and use of illustrative examples of ecological risks and appropriate EAM-based solutions. This could be one way to attract more attention from the national media and thereby reach out to the general public. It could help make environmental issues more of a 'hot topic'. It would also complement initiatives of environmental NGOs and the business sector that address the average citizen in his/her role as consumer (only) (through, for instance, organic consumption or sustainable choice of seafood). To be a responsible consumer is of course one of the most essential roles citizens can play. But more proactive risk communication should also invite and stimulate citizens to discuss, frame and campaign for a cleaner, healthier, more attractive, fair and sustainable Baltic Sea (cf. Dreyer et al. 2011).
3. We also would like to point to the fact that *social scientists* are not used to their full potential (or hardly used at all; cf. Linke and Jentoft 2014) and need to be part of all stages of risk governance. All environmental problems are fundamentally also social problems, whether we are talking about causes, understandings or solutions to problems. Involving social scientists in only select stages of the governance process is likely thus to reproduce a reductionist view and treatment. Social scientists are needed to broaden the perspective and highlight the social character of risk issues. Moreover, using social scientists as mediators in different deliberative processes and meetings could, for example, help by bringing in a common language. NGOs also could take on a greater responsibility for raising awareness on environmental risks in the Baltic Sea area so as to create a common agenda.
4. The precise *form of communication and use of mediums* for communication needs to reflect the transnational nature of risks under consideration, their context and whether they arouse, or could arouse, societal concern. It is important to note that some risks are related to more or less global flows (e.g. chemicals) that even more so accentuate the challenges. Since there is a need to raise the degree of political and public awareness about the five risks in the Baltic Sea region, national *news media* has to highlight cross-country issues. There is a need for common media and shared platforms for debating issues, including regional media stations in different regions (e.g. the Baltic Sea or Mediterranean area).

So far there is no influential transnational (e.g. European) media. Nor is there an international public sphere with a role in governance and political communication. Hence, on the one hand, we have a situation in BSR where an arena for environmental risk governance exists, including regional actors and networks like HELCOM and UBC (and also a request from the European Commission's White Paper on European Governance that a regional or international public should be involved and invited to participate in decision-making procedures). On the other hand, there are obvious obstacles for public deliberation and participation on risk issues in Europe and in different regions in Europe. Important parts of the public sphere are missing and there are no common agendas or arenas for public participation. Such a situation prevents positive development of future governance efforts.

5. Risk governance and communication could also make much more *use of digital media* platforms. Online media offer spaces and platforms where citizens may engage in dialogic communication. So far, ICTs (Internet, social media and the like) are an important source of information for NGOs and other stakeholders who already possess some adequate knowledge about these platforms, but not for the public, who often do not know what to look for and where. Finally, it is important to stress the importance of both multiple and common agendas.

Acknowledgements This chapter draws on results from the research programme 'Environmental Risk Governance of the Baltic Sea' (2009–2015, Michael Gilek, programme coordinator), which involved research teams from Södertörn University in Sweden, Åbo Akademi University in Finland, Dialogik/Stuttgart University in Germany and Gdansk University in Poland. The funding came from the Foundation for Baltic and East European Studies and the European Community's Seventh Framework Programme (2007–2013) under grant agreement no. 217246 made with the joint Baltic Sea research and development programme BONUS, as well as from the German Federal Ministry of Education and Research (BMBF), the Swedish Environmental Protection Agency, the Swedish Research Council Formas, the Polish Ministry of Science and Higher Education and the Academy of Finland. We wish to express our warmest thanks to these institutions for enabling us to conduct this research, to all participants in the research programme that directly or indirectly provided useful inputs, to all informants sharing their experiences and to the two reviewers for their constructive comments on an earlier version of the chapter.

References

Aarhus Convention on Access to Information (1998) Public participation in decision-making and access to justice in environmental matters

Anderson A (1997) Media culture and the environment. Routledge, London

Asp K (1986) Mäktiga massmedier. Studier i politisk opinionsbildning (in Swedish). Akademilitteratur, Stockholm

Baltic 21 (2002) Agenda 21 for the Baltic Sea region sector report – education. Baltic 21. Series No 02/2002

Beck U (1992) Risk society. Sage, London

Boström M, Grönholm S, Hassler B (2016) The ecosystem approach to management in Baltic Sea governance: towards increased reflexivity? In: Gilek M et al (eds) Environmental governance of the Baltic Sea. Springer, Dordrecht

Brake ML, Weitkamp E (2009) Introducing science communication. A practical guide. New York: Palgrave Macmillan

Breakwell GM (2007) The psychology of risk. Cambridge University Press, New York

Bucchi M (2004) Can genetics help us rethink communication? Public communication of science as a 'double helix'. New Gen Soc 23(3):269–283

Bäckstrand K (2003) Civic science for sustainability: reframing the role of experts, policy-makers and citizens in environmental governance. Glob Environ Polit 2003(3):24–41. doi:10.1162/152638003322757916

Castells M (2008) The new public sphere: global civil society, communication networks and global governance. Ann Am Acad Polit Econ Sci 616(1):78–93

Cox R (2010) Environmental communication and the public sphere, 2nd edn. Sage, London

Dahlgren P (2005) The internet, public spheres, and political communication: dispersion and deliberation. Polit Commun 22(2):147–162

De Marchi B (2003) Public participation and risk governance. Sci Public Policy 30(3):171–176

Dreyer M, Sellke P, Boström M, Jönsson AM, Hammer M, Renn O, Söderström S, Zgrundo A (2011) Structures and processes of public and stakeholder communication connected to environmental risks in the Baltic Sea. Deliverable 10 within the RISKGOV project. Available from: http://www.sh.se/riskgov

Dreyer M, Boström M, Jönsson AM (2014) Participatory deliberation, risk governance and management of the marine region in the European Union. J Environ Pol Plan. doi:10.1080/1523908X.2013.866891

European Commission (2001a) Sixth environment action plan 2001–2010. European Commission, Brussels

European Commission (2001b) European governance. A white paper. COM (2001) 428 final. CEC, Brussels

Entman RM (1993) Framing: toward clarification of a fractured paradigm. J Commun 43(4):51–58. doi:10.1111/j.1460-2466.1993.tb01304.x

Egan Sjölander A, Jönsson AM (2012) Contested ethanol dreams – public participation in environmental news. In: Philips L, Carvalho A, Doyle J (eds) Citizen voices, performing public participation in science and environment communication. Intellect Ltd, Chicago, pp 47–71

Felt U, Fochler M (2010) Machineries for making publics: inscribing and de-scribing publics in public engagement. Minerva 48:219–238

Fiske J (1990) Introduction to communication studies, 2nd edn. Routledge, London

Frewer LJ (2003) Trust, transparency, and social context: implications for social amplification of risk. In: Pidgeon N, Kasperson RE, Slovic P (eds) The social amplification of risk. Cambridge University Press, Cambridge

Gilek M, Linke S, Udovyk O, Karlsson M, Lundberg C, Smolarz K, Lemke P (2011) Interactions between risk assessment and risk management for environmental risks in the Baltic Sea. Deliverable 9 within the RISKGOV project. Available from: http://www.sh.se/riskgov

Goffman E (1974) Frame analysis: an essay on the organization of experience. Harper and Row, New York

Gregory J, Miller S (2000) Science in public. Basic Books, Cambridge

Haahti BM, Hedenström E, Linke S, Lundberg C, Reisner G, Wanamo M (2010) Case-study report: eutrophication. Deliverable 2 within the RISKGOV project. Available from: http://www.sh.se/riskgov

Habermas J (1989) The structural transformation of the public sphere. MIT Press, Cambridge

Hagendijk R, Irwin A (2006) Public deliberation and governance: engaging with science and technology in contemporary Europe. Minerva 44:167–184

Hannigan J (2006) Environmental sociology. Taylor and Francis Ltd, London

Hansen A (2010) Environment, media and communication. Routledge, London

Hassler B, Söderström S, Lepola N (2010) Marine oil transportations in the Baltic Sea area. Deliverable 6 within the RISKGOV project. Available from: http://www.sh.se/riskgov

Hassler B, Boström M, Grönholm S (2013) Towards an ecosystem approach to management in regional marine governance? The Baltic Sea context. J Environ Policy Plan 15(2):225–245

HELCOM (2007) HELCOM Baltic Sea action plan. HELCOM Ministerial Meeting. Krakow. 15 Nov 2007

HELCOM (2010) Ecosystem health of the Baltic Sea. HELCOM initial holistic assessment. BSEP No. 122

Irwin A (2008) STS perspectives on scientific governance. In: Hackett E, Amsterdamska O, Lynch M, Wajcman J (eds) The handbook of science and technology studies. MIT Press, Cambridge, MA, pp 82–607

Janse G, Konijnendijk CC (2007) Communication between science, policy and citizens in public participation in urban forestry – experiences from the neighbourwoods project. Urban For Urban Green 6:23–40

Jentoft S, van Son T, Bjørkan M (2007) Marine protected areas: a governance system analysis. Hum Ecol 35:611–622

Joas M, Jahn D, Kern K (2008) Governance in the Baltic Sea region: balancing states, cities and people. In: Joas M, Jahn D, Kern K (eds) Governing a common sea. Environmental policies in the Baltic Sea region. Earthscan, London, pp 3–15

Jönsson AM (2011) Framing environmental risks in the Baltic Sea – a news media analysis. AMBIO 40:121–132

Jönsson AM (2012) Climate governance and virtual public spheres. In: Carvalho A, Peterson R, Amherst T (eds) Climate change politics. Cambria Press, New York

Karlsson M, Gilek M, Udovyk O (2011) Governance of complex socio-environmental risks – the case of hazardous chemicals in the Baltic Sea. AMBIO 40(2):144–157

Klintman M, Boström M (2008) Transparency through labeling? Layers of visibility in environmental risk management. In: Garsten C, Lindh de Montoya M (eds) Transparency in a new global order: unveiling organizational visions. Edward Elgar, Cheltenham

Lemke P, Smolarz K, Zgrundo A, Wolowicz M (2010) Biodiversity with regard to alien species in the Baltic Sea region. Deliverable 4 within the RISKGOV project. Available from: http://www.sh.se/riskgov

Lidskog R, Soneryd L, Uggla Y (2010) Transboundary risk governance. Earthscan, London

Linke S, Jentoft S (2013) A communicative turnaround: shifting the burden of proof in European fisheries governance. Mar Policy 38:337–345

Linke S, Jentoft S (2014) Exploring the phronetic dimension of stakeholders knowledge in EU fisheries governance. Mar Policy 47:153–161

Linke S, Dreyer M, Sellke P (2011) The Regional Advisory Councils: what is their potential to incorporate stakeholder knowledge into fisheries governance? AMBIO 40(2):133–143

Linke S, Gilek M, Karlsson M, Udovyk O (2014) Unravelling science-policy interactions in environmental risk governance of the Baltic Sea: comparing fisheries and eutrophication. J Risk Res 17(4):505–523

Mazzoleni G, Schulz W (1999) "Mediatization" of politics: a challenge for democracy? Polit Commun 16(3):247–261. doi:10.1080/105846099198613

McCombs M (2005) A look at agenda-setting: past, present and future. J Stud 6(4):543–557. doi:10.1080/14616700500250438
McCombs M, Shaw D (1972) The agenda setting function of the mass media. Public Opin Q 36:176–187. doi:10.1086/267990
McGlade J (2010) The main environmental challenges of the 2010's in the Baltic Sea region. Balt Rim Econ 5/2010: 8
Pan Z, Kosicki GM (1993) Framing analysis: an approach to news discourse. Polit Commun 10(1):55–75. doi:10.1080/10584609.1993.9962963
Pidgeon N, Simmons P, Henwood K (2006) Risk, environment, and technology. In: Taylor-Gooby P, Zinn J (eds) Risk in social science. Oxford University Press, New York
Pierre J, Peters GB (2000) Governance, politics and the state. Macmillan, London
Renn O (2008) Risk governance: coping with uncertainty in a complex world. Earthscan, London
Rein M, Schön D (1993) Reframing policy discourse. In: Fischer F, Forester J (eds) The argumentative turn in policy analysis, planning. Duke University Press, London
RISKGOV (2011) Roundtable 3: stakeholder participation and communication in Baltic Sea environmental risk governance. Roundtable discussions with invited stakeholders and experts. Dialogic, Stuttgart. 14–15 Feb 2011. Available from: http://www.sh.se/riskgov
Schulz W (2004) Reconstructing mediatization as an analytical concept. Eur J Commun 19(1):87–101. doi:10.1177/0267323104040696
Schön D, Rein M (1994) Frame reflection. Toward the resolution of intractable policy controversies. Basic Books, New York
Shoemaker P, Reese S (1996) Mediating the message theories of influences on mass media content. Longman, White Plains
Sellke P, Dreyer M, Renn O (2010) Fisheries: a case study of environmental risk governance in the Baltic Sea. Deliverable 3 within the RISKGOV project. Available from: http://www.sh.se/riskgov
Stirling A (2005) Opening up or closing down? Analysis, participation and power in the social appraisal of technology. In: Leach M, Scoones I, Wynne B (eds) Science and citizens. Globalization and the challenge of engagement. Zed Books, London, pp 218–231
Strömbäck J (2008) Four phases of mediatization: an analysis of the mediatization of politics. Int J Press Polit 13(3):228–246. doi:10.1177/1940161208319097
Udovyk O, Rabilloud L, Gilek M, Karlsson M (2010) Hazardous substances: a case study of environmental risk governance in the Baltic Sea region. Deliverable 5 within the RISKGOV project. Available from: http://www.sh.se/riskgov
US NRC (National Research Council) (1989) Improving risk communication. National Academy Press, Washington, DC
Vangas A (2010) Can the EU strategy for the Baltic Sea region bridge the 'great divide'? In: Ozolina Ž, Reinholde I, Rostoks T (eds) EU strategy for the Baltic Sea region. A year after and beyond. Zinatne, pp 103–119
van Deveer SD (2011) Networked Baltic environmental cooperation. J Balt Stud 42(1):37–55
van Leeuwen J, Kern K (2013) The external dimension of European Union marine governance: institutional interplay between the EU and the International Maritime Organization. Glob Environ Polit 13:69–87
Voss JP, Kemp R (2006) Sustainability and reflexive governance: introduction. In: Voss JP, Bauknecht D, Kemp R (eds) Reflexive governance for sustainable development. Edward Elgar Publishing, Northampton
Whiteside KH (2006) Precautionary politics. Principle and practice in confronting environmental risk. The MIT Press, Cambridge
Young O (2009) Governance for sustainable development in a world of rising interdependencies. In: Delmas MA, Young O (eds) Governance for the environment. New perspectives. Cambridge University Press, Cambridge
Zavestoski S, Shulman S, Schlosberg D (2006) Democracy and the environment on the internet electronic citizen participation in regulatory rulemaking. Sci Technol Hum Val 31(4):383–408

Chapter 10
Seeking Pathways Towards Improved Environmental Governance of the Baltic Sea

Michael Gilek and Mikael Karlsson

Abstract Governing marine environments is a highly complex and challenging enterprise. This applies particularly to the heavily exploited Baltic Sea for which despite extensive governance arrangements and a substantial scientific knowledge base, it is unlikely that the policy objective of 'good environmental status' is reached. Based on a review of governance arrangements linked to five large-scale environmental issues (eutrophication, overfishing, invasive alien species, chemical pollution and oil spills from shipping), this chapter aims to identify pathways and concrete ideas for institutional reform that may improve goal fulfilment. The results show that governance challenges differ substantially between environmental issues, implying a need for case-specific management reforms. For example, coping with extreme uncertainty is a key challenge in the chemical pollution case, whereas it seems more pertinent in the eutrophication case to address the complexity of nutrient pollution sources by adapting objectives and measures amongst sectoral policies to be in line with environmental ones. Furthermore, cross-case comparisons reveal a set of common vital functions (i.e. coordination, integration, interdisciplinarity, precaution, deliberation, communication and adaptability) that are needed in order to facilitate effective and efficient environmental governance in the long term. To promote these functions in Baltic Sea environmental governance, the chapter suggests pathways and institutional reforms aimed at improving multilevel and multisectoral integration, science-policy interactions and stakeholder participation. To further develop these ideas, it is proposed amongst other things that priority is given to setting up an international 'Baltic Sea Policy Review Mechanism', formed by cross-body and cross-stakeholder participation.

Keywords Ecosystem approach to management • Marine policy • Environmental policy integration • Science-policy interactions • Stakeholder participation

M. Gilek (✉) • M. Karlsson
School of Natural Sciences, Technology and Environmental Studies,
Södertörn University, 14189 Huddinge, Sweden
e-mail: michael.gilek@sh.se; mikael.karlsson@2050.se

M. Gilek et al. (eds.), *Environmental Governance of the Baltic Sea*,
MARE Publication Series 10, DOI 10.1007/978-3-319-27006-7_10

10.1 Introduction

The aims of this book and the underlying research[1] have been to achieve a better and more comprehensive understanding of the complex structures and processes associated with the governance of the Baltic Sea environment and, based on this, to explore problems and opportunities when trying to cope with the identified key governance challenges (Gilek et al. 2015b). We addressed these aims by characterising the problems and risks and by analysing the governance structures, processes and key challenges associated with five large-scale environmental problems and risks in the Baltic Sea: eutrophication, overfishing, invasive alien species, chemical pollution and oil spills from shipping (Hassler 2016; Karlsson and Gilek 2016; Karlsson et al. 2016; Sellke et al. 2016; Smolarz et al. 2016). Based on these case studies, we subsequently explored the key findings in a cross-case analysis of three important dimensions of primary concern for environmental governance: multilevel and multisectoral structures (Boström et al. 2016), assessment-management processes and interactions (Linke et al. 2016) and stakeholder participation and communication (Jönsson et al. 2016). In each of these eight studies, a number of ideas were already identified on how to potentially develop and improve Baltic Sea governance.

In this final chapter, we attempt to take the case and cross-case conclusions further and seek to identify broader pathways, as well as concrete institutional reforms and strategies that we consider could improve environmental governance structures and processes in the Baltic Sea region (BSR). Clearly, these are formidable tasks, since marine environmental governance often is characterised by multiple and potentially conflicting interests (e.g. fisheries, shipping, recreation and conservation), combined with complex ecosystems and multifaceted governance structures and interactions at local, national and international levels in both the public and private spheres. As a consequence, integrated environmental governance of a regional sea like the Baltic Sea has been considered a 'wicked' problem where problem perceptions amongst stakeholders, sectors and countries usually are contested and management responses mostly less than ideal (cf. Gilek et al. 2015a; Jentoft and Chuenpagdee 2009). However, even though this insight initially led us to adopt a quite 'modest' approach by elaborating ideas for long-term structural and processual reforms based on reflexive thinking (Boström et al. 2016), we believe that the severity and urgency of environmental problems and governance shortcomings in the Baltic Sea is a reason to also attempt to develop proposals for concrete and more directly applicable reform measures. In trying to suggest concrete improvements, we approach it humbly by inviting others to scrutinise and debate our proposals. Hopefully, this will stimulate a constructive process resulting in

[1] This edited volume presents the findings of the research projects RISKGOV (Environmental Risk Governance of the Baltic Sea) and COOP (Cooperating for Sustainable Marine Governance), which were international interdisciplinary research projects focused on understanding practices and challenges for environmental governance of the Baltic Sea. See, e.g. www.sh.se/riskgov

increasingly concrete and well-crafted measures and strategies for improving the governance of the Baltic Sea environment.

Hence, the logic of this concluding chapter is to, based on a summary of key findings in the individual case studies (Sect. 10.2) and cross-case analyses (Sect. 10.3), venture into developing concrete ideas for how environmental governance of the Baltic Sea potentially could be improved based on an identification of 'root problems'. Finally, we summarise key conclusions and recommendations (Sect. 10.4).

10.2 Findings in the Five Individual Case Studies

The five in-depth case studies were identified amongst a set of regional issues that were prioritised in Baltic Sea environmental governance, based on the severity and scope of the associated environmental problems and risks. In Table 10.1, the various problems and risks and their scope, as well as the broad governance patterns, are summarised. As can be seen, the problems are often severe and large scale. Numerous studies have shown that the Baltic Sea is amongst the most disturbed seas worldwide (e.g. HELCOM 2010). In response, the population in the nine countries bordering the Baltic Sea has expressed in monetary terms a willingness to pay nearly 4 billion annually (Baltic Stern 2013) for reducing eutrophication by fulfilling the Baltic Sea Action Plan (HELCOM 2007). Regarding governance patterns, the table shows that the regional level – the EU as well as HELCOM – is nearly always of highest importance, even if local, national and global dimensions play central roles in some of the cases.

Furthermore, it is obvious that the characteristics of the five cases often differ substantially in terms of the complexity of causes and the degree of scientific uncertainty and sociopolitical controversy, as illustrated in Table 10.2. This fact enabled interesting comparisons of governance structures and processes under various conditions.

In general, with the oil case being the main exception, various degrees of more or less high uncertainty and disagreement characterise the cases (Table 10.2). Considering current ambitions to implement the ecosystem approach to management (EAM), implying a need to govern various risks in one and the same ecosystem simultaneously (cf. Boström et al. 2016), the complexity increases even more, due to the various feedback mechanisms involved (remembering also that the impact of climate change will add another complex dimension in the coming decades). In spite of this, a number of governance strategies and tools that can be improved in each of the cases have been identified in the five case study chapters in the book. In the next section, these will be compared and characterised.

In the case of *oil discharges*, it can generally be concluded that much of the needed governance structures and frameworks are in place. IMO acts as a 'global hub', with the EU as a strong enforcer and HELCOM as a catalyst (Hassler 2016). The complexity of sources is comparatively limited and neither uncertainty nor

Table 10.1 Summary of identified environmental problems and risks, scope and governance patterns in the five case studies of environmental governance in the Baltic Sea

	Identified problems and risks	Scope	Governance patterns
Eutrophication[a]	Hypoxia, algae blooms, etc. leading to potentially severe ecosystem disturbances and economic losses	Essentially regional. Different marine sub-regions unequally affected	National governments, EU and HELCOM main actors. Contradictions between CAP and environmental directives
Overfishing[b]	Decreased stocks, disturbances on ecosystems and risk of extinction of stocks. Socioeconomic consequences	Primarily regional, but sub-regional genetic variations cause some local differences	EU, often exclusive, competence. ICES plays important role. RACs attempt to decentralise and improve stakeholder involvement
Invasive alien species[c]	Impact on biodiversity, potentially severe effects on ecosystem levels; economic losses	Global, as ballast water from marine shipping is the main vehicle of entry	Structures have been weak. Recent regulation under implementation. A few stakeholders involved
Chemical pollution[d]	Serious impacts on ecosystems and on human health. Halogenated substances still problematic and several emerging risks	Depends on substance and source; primarily regional but also global product chains. Often most serious effects near the pollution source	Several global conventions, but EU plays the major role. HELCOM important complement
Oil discharges[e]	Large accidents may severely harm ecosystems and socioeconomic interests; operational oil spill constant	Essentially global, as vessels travel globally. Clean-up capability local, national and sub-regional	IMO plays central role as an umbrella for global conventions. HELCOM initiator. EU may strengthen enforcement

Adapted from Hassler et al. (2011)
[a]Karlsson et al. (2016)
[b]Sellke et al. (2016)
[c]Smolarz et al. (2016)
[d]Karlsson and Gilek (2016)
[e]Hassler (2016)

disagreements seem to impede governance to any significant extent. Risk assessment and risk management are relatively straightforward exercises mostly characterised by monitoring and surveillance on the assessment side and a combination of flag and port state controls in terms of management. Creating incentives for key actors has been important, as have been measures taken by proactive states. We consider that continuing along these lines through EU and HELCOM initiatives

Table 10.2 Characteristics of five major environmental problems and risks in the Baltic Sea based on individual case studies (see Stirling (2010) for an elaboration of the concept of uncertainty)

	Complexity of causes	Scientific uncertainty and scientific disagreement	Socio-political controversy
Eutrophication[a]	*High*	*High uncertainty* on ecosystem effects and resilience	*High* among stakeholders, countries, and sectors on prioritisation of management actions and trade-offs among management objectives
	Agriculture, municipalities, traffic, maritime transport, etc.	*Some disagreement* on specific management actions	
Overfishing[b]	*Low*	*High uncertainty*, especially on multi-species management, ecosystem effects and resilience	*Very high* on risk framing (environmental *vs.* socio-economic) and among stakeholders on how to cope with uncertainty
	Commercial fisheries.	*Some disagreement* on risk framing	
Invasive alien species[c]	*Intermediate*	*Extremely high uncertainty* on outcomes of specific new introductions	*Limited* with differences in management priorities among countries, etc.
	Natural and human sources (*e.g.* transports, aquaculture).	*Some disagreement* on risk framing	
Chemical pollution[d]	*High*	*Extremely high uncertainty* on sources, long-term risks and cocktail effects	*High* on how to cope with uncertainty
	Point sources, long-range transport, products, etc.	*Disagreement* on risk evaluation and on how to cope with uncertainty	*Intermediate* on cost-benefit trade-offs and management priorities
Oil discharges[e]	*Low*	*Intermediate uncertainty* on long-term effects, occurrence of intentional discharges and human factor drivers	*Intermediate* on cost-benefit trade-offs and management priorities
	Mainly marine transports (accidental and operational).	*Minor disagreement*	

[a]Karlsson et al. (2016)
[b]Sellke et al. (2016)
[c]Smolarz et al. (2016)
[d]Karlsson and Gilek (2016)
[e]Hassler (2016)

could offer a way to further improve governance. In particular, an increased emphasis on human factors as causes of accidents seems warranted, since human error and performance become more important as other causes are reduced.

Concerning *fisheries*, while the complexity of sources is low, the high uncertainty of some important ecosystem effects in combination with sociopolitical controversies is clearly obstructing governance efforts. One response so far has been to apply a precautionary approach, if not in political decisions on quotas at least in preceding science-based advice. More important are the relatively new arrangements for stakeholder participation (Linke et al. 2016; Sellke et al. 2016). In this case, it seems most important at present to ensure full implementation of the policies in place, which to some extent were recently (2014) renewed in the EU, in order to see if that will be adequate in relation to stated objectives. We consider two dimensions to be particularly important; first, to apply the principle of maximum sustainable yields within the frame of the EAM and the precautionary approach, as well as to phase out discards and subsidies, and, second, to further regionalise decision-making and to improve stakeholder participation.

Similarly, when it comes to *invasive alien species* (IAS) (Smolarz et al. 2016), recent policies have been adopted (EU 2014). While uncertainty in terms of ecosystem effects of IAS is very high, risk management measures, for example, to better control ballast water, seem well founded and relatively unproblematic to implement, as long as international cooperation works smoothly. Still, if an invasive species has high fitness in the Baltic Sea ecosystem, even quite small implementation deficits might cause large problems, in particular over time. Nevertheless, in our view, a critical point seems to be to ensure an ambitious and broad implementation of the new regulation in its three dimensions of prevention, early warning and rapid response and management. Possibly, this could be achieved if, or when, the IMO Ballast Water Management Convention enters into force.

Regarding *chemicals*, it is much more difficult than in the other cases to obtain sufficient knowledge. Present risk assessments, that are afflicted with a number of shortcomings, and cocktail effects, amongst other things, are extremely difficult to evaluate (Karlsson and Gilek 2016). There are several science-based methods for coping with uncertainty, for instance, by using precautionary default values for exposure and toxicity when data is missing and by applying alternative decision-making criteria, such as maximin criteria (Karlsson 2010; Udovyk and Gilek 2013), but present regulatory frameworks in the EU and the nation states around the Baltic Sea have seldom used such approaches (Linke et al. 2016). Improved environmental risk governance in this case would presume vast regulatory reforms in the EU and amongst parties to the Helsinki Convention. We consider it important, first, to fully reverse the burden of proof for decision-making, meaning, for example, that a producer or user of a substance should show that legal requirements for safety are met so that agencies do not have to prove risks beyond a reasonable doubt. Second, regulatory reforms are needed to better coordinate environmental (e.g. the Marine Strategy Framework Directive, MSFD) and polluter-oriented policy approaches (such as the REACH regulation) (cf. Karlsson and Gilek 2016).

Finally, in the case of *eutrophication*, while the basic causes of nutrient leakage are easily identified, the ecosystem and resulting socio-economic effects are far more complex and long-lasting. The difficulty to transform, for example, agricultural production around the Baltic Sea, a dominating source of nutrient leakage, to generally lower levels of nutrient loss, taken together with the strong resistance to do so amongst many farmers and their organisations, makes environmental governance in this case very difficult. This is further complicated by a set of other leakage sources and ambiguity concerning which measures would be most cost effective. Present policies in the EU, HELCOM and individual nation states are far from sufficient to steer development steadily towards agreed targets, which points out a need for both immediate policy-making and longer-term deep reform in the sectors contributing to the problems, as well as in society at large. In the near future, as we see it, pricing externalities in line with the polluter pays principle set out in the EU treaty (meaning, e.g. environmental taxes on fertilisers) and reforming subsidies, steering away from incentivising production not compatible with agreed environmental targets, are examples of potential measures. In the longer run, we consider that deep reforms of agricultural systems might be needed, for example, by improved spatial coordination of crop production and husbandry in order to better control flows of nutrients. At the same time, several of these potential reforms may require an increased willingness to pay amongst consumers for environmental measures in food production.

Evidently, the proposals that we have identified above are not described and evaluated in any detail, and before adopting or implementing such policies and processes, potential consequences should be investigated, whether nationally, in the EU or within HELCOM. We believe though that the ideas presented are motivated to such an extent that they will stimulate discussion and further analysis and studies. In the next section, we zoom out from the specific cases and take a look at governance issues on more of a system level.

10.3 Findings in the Three Cross-Case Studies

As shown in the previous chapters of this book, the *governance structures* in BSR are complex and include formal as well as non-formal components (e.g. Boström et al. 2016). The formal governance consists of institutions and regulatory frameworks at supranational, national and local levels.

At the highest level, both the EU and HELCOM are active in marine governance but have different constellations of members, and while their activities overlap, the policies often have diverging legal strengths[2] and objectives (e.g. concerning improvements in water status) with differences in time plans, approaches (like EAM)

[2] In the EU, binding qualified majority decisions are the normal case, whereas decisions in HELCOM usually presume unanimity and are nonbinding. EU decisions are thus likely to be implemented nationally to a much greater extent than decisions under the Helsinki Convention.

and measures for implementation. The international policies in place also span different sectors, but the mechanisms for coordinating them vary and are far from sufficiently developed as, for example, clearly illustrated in the EU Strategy for BSR (EUSBSR) and the HELCOM Baltic Sea Action Plan (BSAP) relating to, for example, eutrophication (Karlsson et al. 2016). A parallel situation with sectors' cleavages and tensions (e.g. between environmental protection on the one hand and the use of natural resources on the other) often exists at the national and local levels. The prospects for radical multilevel and multisectoral coordination and collaboration in the near future are therefore rather small, but it should not be forgotten that vertical and horizontal interactions in some situations take place by 'uploading' HELCOM recommendations into binding EU directives (Gilek et al. 2015a).

Adding to this complexity, the governance institutions and processes have developed rapidly over the years (Boström et al. 2016; Jönsson et al. 2016), recently by including venues for stakeholder participation such as Regional Advisory Councils (RACs) in EU fisheries management (Sellke et al. 2016). What has also developed rapidly is the extent to which nations in the region have put efforts into marine environmental governance, spanning from forerunners to those whose activities were limited until EU membership, with the exception of those who are still lagging behind on implementation.

Moreover, numerous actors and networks operate in non-formal governance structures in the region, carrying out countless projects in the marine governance field (Boström et al. 2016). All in all, the number of possible interactions, both vertically and horizontally, is massive, which not only opens up the possibilities for collaboration and learning, for instance, between sectors (e.g. HELCOM Fisheries/Agriculture Forums in relation to BSAP implementation) but may also in other contexts impede possibilities to steer developments and bridge various sector interests. Hence, despite dense and highly interactive multilevel and multi-actor governance structures, integration between these is commonly insufficiently developed.

Regarding *assessment-management interactions*, the Baltic Sea is often referred to as one of the best-investigated seas in the world, which has laid a foundation for generating science-based advice (e.g. HELCOM 2010; Udovyk and Gilek 2013). This has in some instances led to, at least partially, successful management measures, as seen, for example, in HELCOM's identification and management of pollution hotspots and some hazardous chemicals such as PCBs, despite long recovery times from such marine pollution (Karlsson et al. 2011; Karlsson and Gilek 2016).

A mismatch often exists between the more common regional scientific assessments and the frequently used national management strategies and measures (Linke et al. 2016). Moreover, even in situations when assessment and management regimes address the same level, they often focus on diverging policies and organisational requirements (e.g. EU MSFD and HELCOM BSAP), without sufficient coordination (Karlsson et al. 2016). These institutional and other mismatches cause gaps and overlaps between assessment and management, as well as in the operational chain spanning from definition of environmental objectives over environmental assessment and monitoring to implementation of management measures.

This means that despite some successful exceptions as exemplified above, science-based advice is far from always used effectively in Baltic Sea environmental governance. This is apparent in the cases of eutrophication (Karlsson et al. 2016) and chemicals (Karlsson and Gilek 2016), where HELCOM has established detailed regional assessments based on scientific input, but management measures are nonetheless seldom fully implemented nationally. Furthermore, risk assessments are usually established based on a conventional view of what constitutes appropriate scientific methodologies and knowledge, often overlooking non-standardised data sources, uncertainty and interactions between various risks, as well as the need for interdisciplinary perspectives and stakeholder input (Linke et al. 2016). The latter – lack of stakeholder input – might cause worsened sociopolitical controversies, especially in the presence of uncertainty. In particular, it opens up for strong politicisation where scientists without normative transparency engage in political discussions and politicians selectively interpret scientific results (Karlsson et al. 2011; Linke et al. 2016).

Several chapters in the book analyse and discuss *communication and stakeholder participation*. Both the EU and HELCOM have invested increasingly in this area of environmental governance in recent years, as, for example, seen in HELCOM's BSAP stakeholder conferences and in RACs under the Common Fisheries Policy (Boström et al. 2016; Jönsson et al. 2016; Sellke et al. 2016). To a large extent though, the case studies reveal that participation in Baltic Sea environmental governance is generally regarded as having an instrumental role to serve the requirements of public policy (Boström et al. 2016). It is of course positive if participation, as assumed in this instrumental framing, leads to more efficient and effective environmental governance and higher acceptance of decision-making processes. Still, this instrumental focus on participation may result in the broader democratic values of participation being overlooked (Jönsson et al. 2016). In addition, our findings indicate that regional structures and processes for stakeholder input and critique are often undeveloped or missing, as seen in the chemicals and IAS cases (Linke et al. 2016; Smolarz et al. 2016). Hence, despite ambitions to develop participation in environmental governance and recent developments of, for example, the RAC system in fisheries management, it can be concluded that regional structures and processes for stakeholder inclusion and deliberation generally remain rather undeveloped in BSR.

Finally, there is an obvious lack of widely available supranational communication arenas in the Baltic Sea region, such as a common Baltic news media, which undermines effective environmental communication. Media coverage at the national level, on the other hand, is much more prominent, often making international coverage invisible (Jönsson et al. 2016). This is likely to obstruct opportunities for environmental governance of the Baltic Sea, since the possibility of stakeholders participating in regional societal debates is limited, as is the potential to develop a common regional understanding of environmental challenges and opportunities. In Table 10.3, we summarise the problems we have found to be important in the cross-case analysis, identify specific problem areas and give concrete examples that illuminate our findings better.

10.3.1 Conclusions Based on the Three Cross-Case Studies

In spite of high policy ambitions, many initiatives and efforts made by a wide set of actors and stakeholders, the overall conclusion of the three cross-case studies is that implementation and enforcement generally lag behind in relation to existing objectives for the Baltic marine environment and that this to a significant extent is associated with 'imperfections' in the studied governance structures and processes as outlined above. However, to address the aim of this chapter – to develop more concrete ideas for improvements – it is important to ask whether it is possible to identify any root problems and causes for these implementation deficits.

Based on the specific problems and shortcomings identified in the cross-case comparisons of the focused governance dimensions (multilevel and multisectoral structures, assessment-management interactions and stakeholder communication and participation), it is possible to discern a set of recurring problem areas (Table 10.3). These problem areas have in previous governance research been identified as key governance challenges (Söderström et al. 2015) and have for the purpose of our analysis, given its limitations, been classified as 'root' problems.

In the following discussion, these identified root problems – together with conclusions from the individual cases – provide a basis to formulate broader pathways as well as associated specific ideas about measures and 'institutional reforms' to potentially improve Baltic Sea environmental governance (Table 10.4). In general, we conclude that it is difficult to go much further with regard to adopted environmental targets for the Baltic Sea, without more fundamental changes, i.e. efforts for improvement must consist of something else than 'more of the same'. This is challenging and complex and requires a continuous and adaptive policy-making and transition process. However, despite these challenges, some positive steps have already been taken in line with our proposed pathways, albeit often in rudimentary ways or only in specific cases. Hence, despite difficulties, we do not see the proposed pathways as impossible to embark on more broadly.

It is hardly surprising, looking at the three cross-case governance dimensions focused on in the book, that the root problems differ between governance structures, assessment-management interactions and stakeholder participation (Table 10.4). Still, even though root problems such as 'insufficient coordination and integration' and 'insufficient flexibility and adaptability' were most influential and problematic in the case of governance structures, these problems are also significant and important with regard to other governance dimensions. This means that Table 10.4 should not be seen as an attempt to strictly differentiate between totally different root problems and pathways for the studied governance dimensions. Instead, the table is an attempt to organise our analysis by highlighting key root problems and potential pathways associated with the studied governance dimensions.

This analysis reveals that current *multilevel and multisectoral governance structures* mainly are hampered by insufficient coordination and integration, as well as insufficient flexibility and adaptability (Table 10.4). In our cross-case analysis, we identified a set of specific ideas that together can promote a pathway for improved

Table 10.3 Illustrative root problems and specific examples of problems in the governance of the Baltic Sea environment

Root problem	Identified problem areas and specific examples
Insufficient coordination and integration	*Lack of multi-level coordination:* e.g. diverging policy objectives, time frames and measures in the EU, HELCOM and individual countries as seen e.g. for eutrophication (Karlsson et al. 2016)
	Lack of multi-sector coordination: e.g. often limited integration of sector policies for agriculture, chemicals, fisheries etc. with environmental policies based on EAM, such as the EU MSFD (Karlsson and Gilek 2016; Linke et al. 2016; Sellke et al. 2016)
	Incongruent assessment-management structures: e.g. often a spatial mismatch between regional level assessments and national or European level management, as seen e.g. in eutrophication and chemicals management (Karlsson and Gilek 2016; Karlsson et al. 2016)
	Knowledge deficits: e.g. common lack of interdisciplinary assessments and socio-economic appraisals, undeveloped integration of practitioner knowledge, ignorance and uncertainty in some cases (e.g. Linke et al. 2016; Smolarz et al. 2016)
Insufficient flexibility and adaptability	*Path dependency and lock-in:* e.g. the TAC machine in fisheries policy (Sellke et al. 2016; Linke et al. 2016) and the burden of proof requirements in risk assessment of chemicals (Karlsson and Gilek 2016)
	Non-adaptive governance structures: e.g. time-consuming decision-making arenas (28 Member States need to coordinate within the EU; HELCOM requires consensus to be effective) or processes, such as strong analysis requirements on agencies and multiple bodies in chemicals policy (Karlsson and Gilek 2016).
	Enforcement possibilities: e.g. non-binding decisions in the case of HELCOM policies (Hassler 2016; Smolarz et al. 2016), long-range transportation of pollutants (Hassler 2016) or organisms (Smolarz et al. 2016)
Insufficient coping with uncertainty	*Lack of data and knowledge:* e.g. vast shortcomings of the data required by regulations for managing chemicals (Karlsson and Gilek 2016), ignorance of interactions between species in fisheries management and between different risk areas, and uncertainty of long-term effects of IAS (Smolarz et al. 2016)
	Inadequate regulatory requirements: burden of proof commonly placed on the risk recipient side and not on the polluter or operator (Karlsson and Gilek 2016); cost-benefit-weighing of measures in spite of the existence of skewed data
	Inability to make decisions under uncertainty: e.g. unawareness and lack of regulatory support for applying decision-making tools and criteria for coping with uncertainty, e.g. precautionary default values and substitution-based strategies (Karlsson and Gilek 2016)
Undeveloped stakeholder inclusion and deliberation	*Instrumental framing of participation:* e.g. in RACs under the EU fisheries policy (Sellke et al. 2016)
	Role of participation in EAM: e.g. that participation is not seen as equally important as science for implementing the EAM (Jönsson et al. 2016)
	Neglect of stakeholder values and critique: e.g. a general lack of inclusive arenas for stakeholder input and critique (Boström et al. 2016)
	Lack of regional-level public communication and discourse: e.g. there are only a few examples of forums and media that allow or are open for public communication and debate at the regional Baltic Sea level (Jönsson et al. 2016)

Table 10.4 Identified pathways towards improved environmental governance of the Baltic Sea. Specific ideas for how to promote pathways as well as institutional reform are also indicated (these are further discussed in the text)

	Multi-level and multi-sector governance structures	Assessment – management interactions	Stakeholder participation and communication
Root problems	Insufficient coordination and integration	Insufficient coordination and integration	Undeveloped stakeholder inclusion and deliberation
	Insufficient flexibility and adaptability	Insufficient coping with uncertainty	
Identified pathways	'Towards regionally integrated and reflexive governance arrangements'	'Towards post-normal[a] science based advice and precautionary strategies'	'Towards inclusive stakeholder deliberation'
Specific ideas for promoting the identified broad pathways	Develop existing rudimentary synergies between the EU (e.g. MSFD) and HELCOM's (e.g. BSAP) environmental policies, for example by synthesising BSAP and EUSBSR. Enforcement can often be improved by rescaling regional initiatives to EU regulations	Explicit requirements for interdisciplinary, socio-economic (incl. cost of no action) assessments, as well as stakeholder and practitioner input	Make provisions for stakeholder inclusion and deliberation more explicit in EAM implementation, e.g. linked to implementation of the EU MSFD, the HELCOM BSAP and the EUSBSR
	Reform sector policies (e.g. CAP, CFP, REACH) to strengthen interactions with environmental policies (e.g. MSFD, BSAP). Integrative policies such as MSP[b] can be important mechanisms	Explicit requirements for uncertainty appraisal and development of regionally common guidelines for this	Enhanced efforts to communicate regional level environmental issues and governance challenges, as well as environmental values, services etc. to the general public
	Make explicit requirements for continuous review and reform of governance arrangements based on key criteria such as participation, precaution, polluter pays, adaptive learning and equity	Regulatory provisions for changed burden of proof and other types of precautionary measures	Institutionalise forums and media for generating a stronger Baltic identity, seeking to ensure that Russia and Russian stakeholders are also stimulated to participate
		Institutional reforms to improve the regional and multi-sector basis for integrated science-based advice; streamlining the time-consuming system of analysing and decision-making bodies	

(continued)

Table 10.4 (continued)

	Multi-level and multi-sector governance structures	Assessment – management interactions	Stakeholder participation and communication
Associated ideas for institutional reform to develop required governance functions	*Baltic Sea Policy Review Mechanism*	*Baltic Sea Science Panel*	*Regional Marine Advisory Panel*
	For recurring review and reflection on multi-level, multi-sector and multi-actor governance arrangements	To serve as a regional interdisciplinary source for assessments, science-based advice and guidelines, e.g. as an interdisciplinary regional section under ICES	To support e.g. BSAP, MSFD, MSP with stakeholder advice – e.g. organised in sectoral sections and a cross-sectoral forum

[a]Post-normal science builds on the acknowledgement of fundamental uncertainties and integration of interdisciplinary and stakeholder knowledge and has been proposed as a necessary form of science-based advice on complex environmental issues (e.g. Funtowicz and Ravetz 1993)

[b]Marine Spatial Planning

environmental governance. Taken together we see possibilities that these measures and reforms could promote a pathway *towards regionally integrated and reflexive governance arrangements.*

First, based on previous literature, it is known that there often is a synergetic relationship between the processes of Europeanisation and regionalisation (e.g. in the Baltic Sea region) in marine environmental policy (Gilek et al. 2015a). This can, for example, be illustrated by the mutually reinforcing relationships between the EU MSFD and HELCOM BSAP (Gilek et al. 2015b). We argue that these synergistic multilevel relationships can be strengthened further by coordinating implementation, but moreover by actually adjusting each of these policy schemes so that they address gaps and ineffective overlaps. In addition, coordination is needed with the EU Strategy for BSR, which ideally could serve as a bridging instrument. It would also be highly beneficial to, as far as needed and possible, attempt to 'rescale' regional initiatives from, for example, HELCOM, and make them into binding EU directives or regulations as a means to improve enforcement possibilities in the EU members states around the Baltic Sea.

Second, to improve possibilities for multisectoral coordination and integration, we see substantial possibilities to reform sectoral policies such as CAP, CFP and REACH in order to strengthen their interactions with environmental policies, such as WFD, MSFD and BSAP. Without overlooking strong stakeholder interests striving to preserve as much control over policies as possible, we argue that coordination would be more of a win-win exercise than non-coordination, since present sector policies allow or even subsidise a development that society then tries to govern by imposing environmental policies in terms of laws and taxation. Basically, such insufficient multisectoral coordination creates a situation of conflictual incentives and suboptimal measures for farmers, fishermen, etc., without any long-term safety from either an environmental or market point of view. Well-coordinated frameworks would be more rewarding and easier to deal with from multiple, including environmental and economic, points of views. Here, we see the current ambitions and initiatives to develop integrative policies in the form of marine spatial planning as a potentially important step to improve multisectoral integration in the governance of Baltic marine territory and resources (e.g. Gilek et al. 2015a). However, MSP is at an early stage of development, especially in relation to transboundary governance challenges, such as in the case of the Baltic Sea, which means that substantial efforts are needed in terms of both research and practice in the coming years to develop ideas, processes and approaches that could facilitate integrative MSP.

Finally, we argue that marine governance always will be a work in progress, not least considering the commonly evolving character of natural systems and factors such as policy aims, environmental status and values, stakeholder interests and stakes (cf. Gilek et al. 2015a). This means that marine governance arrangements and aims will continuously need to be reviewed and reformed in a reflexive manner to adapt to new contexts and challenges. In order to do so, we argue that there is a need to set up an institution of one type or another to regularly review and reform Baltic environmental governance – a 'Baltic Sea Policy Review Mechanism' (Table 10.4). It should be further investigated how this 'mechanism' could be achieved – for example, if it should be part of existing institutions or not, if it should be a temporary or standing body and if governments should play a role themselves or rather appoint a more independent top-level forum. In the further development of this mechanism, models in other areas could be analysed, for instance, the GOC on oceans, the IPCP on chemicals, the IPCC on climate and the IPBES on biodiversity, which all have different aims, compositions, functions and ways of operation.[3]

Regarding *assessment-management interactions*, we have concluded that there are insufficient coordination and integration and insufficient handling of uncertainty. We have also identified a need to acknowledge various forms of incertitude (Table 10.4; cf. Linke et al. 2016), for example, by applying interdisciplinary assessment and management approaches and methods from post-normal science studies, as well as science-based precautionary management strategies.

First, it is important to set up assessment strategies that support and develop interdisciplinary approaches and incorporate laymen's and stakeholder's practical knowledge. Besides improved natural science data and studies, knowledge directly needed from a management point of view is considered important, not least in terms of socio-economic data on the so-called 'cost of no action' (the Baltic Stern project is a good start in this respect, cf. Baltic Stern 2013).

Second, explicit regional requirements and guidelines for uncertainty appraisal need to be developed. In addition, regulatory provisions are needed to cope with uncertainty by changing the burden of proof and imposing other types of precautionary measures (cf. Karlsson 2005; Udovyk and Gilek 2013). One important measure would be to allow science-based precautionary default values when data is

[3] See http://www.globaloceancommission.org/about-the-commission/mandate/, http://www.ipcp.ch/about-ipcp, http://www.ipcc.ch/organization/organization.shtml, http://www.ipbes.net/about-ipbes.html

missing, for example, by classifying substances in groups according to so-called worst-case assumptions or by assuming that exotic species are always invasive unless scientific studies reasonably indicate the opposite (Cooney and Dickson 2005; Karlsson 2010; Sandin and Hansson 2002). In contrast with the common decision-making approach to weigh costs and benefits, there are good reasons to instead, or as a complement, apply the maximin criteria to minimise the probability of the worst-case scenario, since data on costs and benefits often are missing or uncertain (cf. Hansson 1997). In some cases, this has to be institutionalised as hard regulation, but in other cases soft policy and regulatory approaches might be possible and even preferable as a testing ground where proactive stakeholders can show a way forward that others can eventually follow. A combination of soft and hard regulations can often be rewarding (cf. Hassler 2016).

Finally, a smoother transfer of data and knowledge from assessment to management is needed, hand in hand with a more holistic approach in the design of decision-making bodies. This relates to improved sectoral integration of science-based advice (e.g. eutrophication and fisheries are interrelated in numerous ways, such as oxygen depletion affecting the survival of cod eggs) and the need for more streamlined management systems. Examples of the latter are the multiple and time-consuming processes of integrating scientific data on hazardous chemicals in the REACH regulatory system, where long-lasting negotiations and interpretation exercises have often replaced an efficient use of new scientific evidence and where arbitrary and normative thresholds place an unreasonably high burden of proof on agencies before decisions can be made (cf. Karlsson 2010).

In order to accomplish these various points in a coordinated and rational manner, we see a need for what could be called a 'Baltic Sea Science Panel', which potentially could be developed as part of the International Council for the Exploration of the Sea (ICES) (Table 10.4).

In the area of *stakeholder participation and communication*, our critique is that there is an underdeveloped situation in which participation is framed instrumentally. Also there are recurring problems of representation and power (Table 10.4; Boström et al. 2016). In response to this, we have identified several possible initiatives that together could pave the way for a pathway *towards inclusive stakeholder deliberation*.

First, it is important to make provisions for stakeholder inclusion and deliberation more explicit in EAM implementation, for example, linked to implementation of the EU MSFD, HELCOM BSAP and EUSBSR. Despite recognition of the fundamental role of stakeholder input in the so-called Malawi principles for an ecosystem approach (cf. Hammer 2015), EAM in the Baltic Sea is today primarily framed as being based on the best available scientific knowledge. This is, for example, seen in the HELCOM definition of EAM (cf. Karlsson et al. 2016). We believe that there are strong instrumental (e.g. linked to governability and governance outcomes) and normative arguments (e.g. linked to democratic ideals of just representation) for striving to complement this science-based approach with a stronger focus on developing participatory aspects of EAM (cf. Jönsson et al. 2016). This could substantially improve possibilities for improved stakeholder input and advice in the governance of the Baltic marine environment.

Second, we see a substantial need for enhanced efforts to communicate environmental issues and governance challenges, as well as environmental values, services, etc., to the general public in BSR. As an example, Jönsson et al. (2016) mention that environmental communication could be prioritised by HELCOM and national authorities as part of BSAP implementation and that this subsequently could attract the attention of media and thereby reach out to the general public. This could complement and even enhance communication efforts by other actors such as business sector organisations and environmental NGOs. All in all, enhanced regional level environmental communication could turn the Baltic Sea environment into a hopefully somewhat 'hotter' topic in regional public debate and ultimately stimulate wider engagement to participate in proactive discussions on environmental governance (cf. Jönsson et al. 2016).

Finally, there is a need to set up regional forums for stakeholder advice, as well as regional media and communication platforms for generating a stronger Baltic identity that include Russia and Russian stakeholders. It seems unrealistic today to develop an 'all-inclusive' institution for stakeholder advice that involves all stakeholders from all sectors and that integrates stakeholder opinions and critiques of all relevant policy areas. However, we believe that a 'Regional Marine Advisory Panel', supporting, for example, BSAP, MSFD and MSP with stakeholder advice, could be set up by combing sectoral subdivisions with integrating forums (cf. Dreyer and Sellke 2015).

10.4 Concluding Remarks

These identified pathways towards improved environmental governance of the Baltic Sea are in need of further analysis and consideration, not least when it comes to how they potentially could interact with each other. Similarly, while a complex reality might seem to call for complex governance structures and processes, overlaps, gaps and counteracting policies are seldom fruitful, and hence we want to caution against creating even more of a governance thicket than today. In this respect, some of our proposals to reform bodies might seem counterproductive. However, while these bodies could fit in or replace current institutions, we want to underline that the seven identified *functions* – coordination, integration, interdisciplinarity, precaution, deliberation, communication and adaptability – will continue to be the most important aspects that need to be taken into account. Whichever governance set-up that is chosen, these aspects cannot be overlooked, as they largely are today, when striving towards improved governance of the Baltic Sea environment and its natural resources. Undoubtedly, further investigations would be needed on how to structure these – or similar – coordinating bodies so that they really promote the vital governance *functions* that are strikingly missing or underdeveloped today. To further develop these ideas, we suggest that priority is given to setting up the proposed international 'Baltic Sea Policy Review Mechanism' that can be formed by cross-body and cross-stakeholder participation. Whether or not this specific proposal will be realised is less important than the need for fundamental reforms based on the functions and ideas discussed here and in the other chapters of this book if improved environmental governance of the Baltic Sea is to be realised.

Acknowledgements This chapter is based on research undertaken within the research projects RISKGOV 'Environmental Risk Governance of the Baltic Sea' (2009–2015) and COOP 'Cooperating for Sustainable Regional Marine Governance'. RISKGOV involved research teams from Södertörn University in Sweden, Åbo Akademi University in Finland, Dialogik/Stuttgart University in Germany and Gdansk University in Poland. Funding came from the Foundation for Baltic and East European Studies and the European Community's Seventh Framework Programme (2007–2013) under grant agreement no. 217246 made with the joint Baltic Sea research and development programme BONUS, as well as from the German Federal Ministry of Education and Research (BMBF), the Swedish Environmental Protection Agency, the Swedish Research Council FORMAS, the Polish Ministry of Science and Higher Education and the Academy of Finland. We thank these institutions for enabling us to conduct this research. We also thank all participants in the research programmes for providing valuable input.

References

Baltic Stern, SWAM (Swedish Agency for Marine and Water Management) (2013) The Baltic Sea – our common treasure. Economics of saving the sea. Rapport 2013:4, SWAM

Boström M, Grönholm S, Hassler B (2016) The ecosystem approach to management in Baltic Sea governance: towards increased reflexivity? In: Gilek M et al (eds) Environmental governance of the Baltic Sea. Springer, Dordrecht

Cooney R, Dickson B (eds) (2005) Biodiversity and the precautionary principle. Risk and uncertainty in conservation and sustainable use. Earthscan, London

Dreyer M, Sellke P (2015) The Regional Advisory Councils in European fisheries: an appropriate approach to stakeholder involvement in an EU integrated marine governance? In: Gilek M, Kern K (eds) Governing Europe's marine environment. Europeanization of regional seas or regionalization of EU policies? Ashgate Publishing, Farnham

EU (2014) Regulation (EU) 1143/2014 of the European Parliament and of the Council on the prevention and management of the introduction and spread of invasive alien species. OJ L 317:35–55

Funtowicz S, Ravetz JR (1993) Science for the post-normal age. Futures 25:739–755

Gilek M, Hassler B, Jentoft S (2015a) Marine governance in Europe: problems and opportunities. In: Gilek M, Kern K (eds) Governing Europe's marine environment. Europeanization of regional seas or regionalization of EU policies? Ashgate Publishing, Farnham

Gilek M, Karlsson M, Udovyk O, Linke S (2015b) Science and policy in the governance of Europe's marine environment – the impact of Europeanization, regionalization and the ecosystem approach to management. In: Gilek M, Kern K (eds) Governing Europe's marine environment. Europeanization of regional seas or regionalization of EU policies? Ashgate Publishing, Farnham

Hammer M (2015) The ecosystem management approach. Implications for marine governance. In: Gilek M, Kern K (eds) Governing Europe's marine environment. Europeanization of regional seas or regionalization of EU policies? Ashgate Publishing, Farnham

Hansson SO (1997) The limits of precaution. Found Sci 2:293–306

Hassler B (2016) Oil spills from shipping: a case study of the governance of accidental hazards and intentional pollution in the Baltic Sea. In: Gilek M et al (eds) Environmental governance of the Baltic Sea. Springer, Dordrecht

Hassler B, Boström M, Grönholm S, Kern K (2011) Environmental risk governance in the Baltic Sea – a comparison among five key areas. RISKGOV report, delivery number 8. Södertörn University, Huddinge, Available from: http://www.sh.se/riskgov

HELCOM (2007) Baltic Sea Action Plan. Adopted at HELCOM Ministerial Meeting, Krakow 15/11/07

HELCOM (2010) Ecosystem health of the Baltic Sea 2003-2007: HELCOM initial holistic assessment. Baltic Sea Environmental Proceedings No. 122

Jentoft S, Chuenpagdee R (2009) Fisheries and coastal governance as a wicked problem. Mar Policy 33:553–560

Jönsson AM, Boström M, Dreyer M, Söderström S (2016) Risk communication and the role of the public: towards inclusive environmental governance of the Baltic Sea? In: Gilek M et al (eds) Environmental governance of the Baltic Sea. Springer, Dordrecht

Karlsson M (2005) Managing complex environmental risks for sustainable development: policies for hazardous chemicals and genetically modified organisms. Doctoral thesis, Karlstad University Studies 2005:34. Karlstad University

Karlsson M (2010) The precautionary principle in EU and US chemicals policy: a comparison of industrial chemicals legislation. In: Eriksson J et al (eds) Regulating chemical risks: European and global challenges. Springer, Dordrecht

Karlsson M, Gilek M (2016) Governance of chemicals in the Baltic Sea region: a study of three generations of hazardous substances. In: Gilek M et al (eds) Environmental governance of the Baltic Sea. Springer, Dordrecht

Karlsson M, Gilek M, Udovyk O (2011) Governance of complex socio-environmental risks: the case of hazardous chemicals in the Baltic Sea. Ambio 40:144–157

Karlsson M, Gilek M, Lundberg C (2016) Eutrophication and the ecosystem approach to management: a case study of Baltic Sea environmental governance. In: Gilek M et al (eds) Environmental governance of the Baltic Sea. Springer, Dordrecht

Linke S, Gilek M, Karlsson M (2016) Science-policy interfaces in Baltic Sea environmental governance: towards regional cooperation and management of uncertainty? In: Gilek M et al (eds) Environmental governance of the Baltic Sea. Springer, Dordrecht

Sandin P, Hansson SO (2002) The default value approach to the precautionary principle. Hum Ecol Risk Assess 8:463–471

Sellke P, Dreyer M, Linke S (2016) Fisheries: a case study of Baltic Sea environmental governance. In: Gilek M et al (eds) Environmental governance of the Baltic Sea. Springer, Dordrecht

Smolarz K, Biskup P, Zgrundo A (2016) Biological invasions: a case study of Baltic Sea environmental governance. In: Gilek M et al (eds) Environmental governance of the Baltic Sea. Springer, Dordrecht

Söderström S, Kern K, Boström M, Gilek M (2015) Environmental governance and ecosystem management: avenues for synergies between two approaches. Interdiscip Environ Rev (In press)

Stirling A (2010) Keep it complex. Nature 468:1029–1031

Udovyk O, Gilek M (2013) Coping with uncertainties in science-based advice informing environmental management of the Baltic Sea. Environ Sci Policy 29:12–23

Index

M. Gilek et al. (eds.), *Environmental Governance of the Baltic Sea*, MARE Publication Series 10, DOI 10.1007/978-3-319-27006-7